Micro Eco-Farming

Prospering from Backyard to Small Acreage In Partnership with the Earth

Barbara Berst Adams

New World Publishing
Auburn, California

Micro Eco-Farming:

Prospering from Backyard to Small Acreage in Partnership with the Earth

By Barbara Berst Adams

Every effort has been made to make this book as accurate as possible and to educate the reader. This text should be used only as a guide, however, and not as an ultimate source of small farm or market gardening information. The author and publisher disclaim any liability, loss or risk, personal or otherwise, which is incurred as a consequence, directly or indirectly, of the use and application of any of the contents of this book.

Cover painting: Randy Griffis, North San Juan, California

Library of Congress Cataloging-in-Publication Data

Adams, Barbara Berst, 1957-

Micro eco-farming : prospering from backyard to small acreage in partnership with the earth / Barbara Berst Adams.

p. cm.

Includes index.

ISBN 0-9632814-3-7 (pbk.)

1. Organic gardening. I. Title.

SB453.5.A32 2004

635'.0484--dc22

2004005128

New World Publishing
11543 Quartz Dr. #1
Auburn, CA 95602
www.nwpub.net

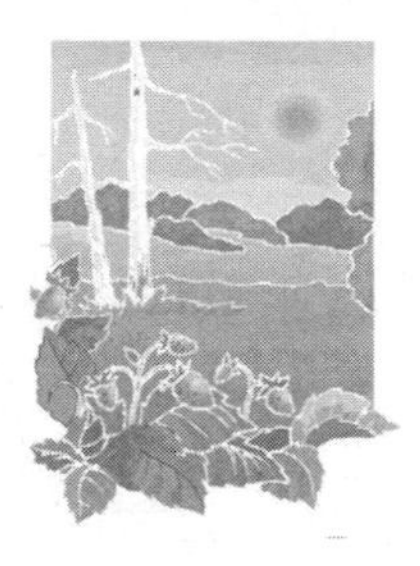

In this timely and informative book, Barbara Berst Adams explores a quiet revolution: the emergence of "microfarms" on tiny acreages that achieve astounding yields of organic produce and provide a better-tasting alternative to the products of big agri-business. Anyone interested in microfarming, food health, the environment, or the future of agriculture will learn from the success stories and lessons in this lively account.

—William Dietrich, Pulitzer Prize winning environmental journalist and author of *Natural Grace* and *Northwest Passage*

Dedication

For my parents, Buddy and Johnnie Berst

Acknowledgements

Thanks begin with my ancestors: the Adams, Prossers, Johnsons, Dinneens, Rosens, and others—the writers and farmers; presidents and first ladies; Quakers and trail blazers—who inspired the roads I chose. Next, thanks to my mother and father, Johnnie and Buddy Berst, who gave me a wonderful childhood full of nature, art and writing. Thank you to my brother and sister, Jesse Berst, Janine Lawless and their families, always there to support every turn my life took. Thank you, my daughter and son, Elise (the event planner who reads the human heart like magic) and Jereme (the webmaster and designer with the gift for deeply embracing both the language of technology and the language of humanity's soul at the same time), for letting me raise you on the farm and watch you climb the big cherry tree, build tree forts, sell pumpkins, ride ponies, rescue baby birds, pick berries, sell flowers, and let your pet goslings take bites from your homemade popsicles. You grew into beautiful adults who never stopped your support and belief and are in me forever and in all that I do. Much appreciation goes to my cousins Don Clarkson, the excellent professional and personal coach, and Vicki Shaffer for her ongoing encouragement. Thank you to Caily Rose, for collecting goose eggs and feeding the bunnies and ducks and horses, and bringing your spirit to the farm. Thanks to all the neighbors on Sharpe Road, always there in times of need, and the town of Anacortes and its writers' group for being an inspiration. Thanks to good friends Lois Lenz, Sharon Mann, Amy Mazza, and Bill Gluth for your shining support and caring. So much gratitude goes to all the incredible farmers of this world and the experts and teachers who took time out of their busy lives to share with me their expertise, advice, secrets and even the deepest part of their hearts for this book's creation. Thank you to my publisher, who most definitely contributes toward a better new world. Thanks to Royal Adams for believing in this project and helping make it possible. Everything you touch blossoms, Royal. And finally, thank you, Kipp Davis, for your love, protection and support.

Publisher's Note: While we believe this book will prove to be of enduring value as a guide for small farmers and market gardeners, no sensible gardening or farming writer would pretend that this ancient art can be mastered from a book. This book is about possibilities and a new horizon. Every farm's bio-region and community is unique. The art and science of farming and marketing evolve continually, with new techniques appearing daily. You will have to create your own particular goals and visions, and then find the specific how-tos for the segments of the farm you choose to weave into a new whole, many of which may be local and often unwritten, which is part of the ongoing art of farming. See our Resources section for some great places to start. This applies especially when it comes to legal matters. Many examples are given in this book in the food handling and composting arena, for example, where strict laws change and are different for different regions. Some areas for example, don't allow animals to graze under orchards for up to a month before fruit harvest. The reader must take responsibility for checking out laws in his or her area. Also, the term "organic" is used in this book as a growing term, rather than a legal term (as it was meant before the government took it over). You will need to check closely about the organic laws in your area before labeling or calling your products "organic."

Contents

Foreword

Barbara Berst Adams' book is all about why "small" matters, and as I write this on a train on the way home to New York City from a trip to rural Pennsylvania, I can't stop thinking how right she is.

I'd come to Pennsylvania with GreenTreks Network (the Emmy Award winning television, radio and networking company) to cover a story on the emergence of "greenways" across the state—a project providing safe, beautiful, protected trails for people of all ages (I met an 87-year-old who still rides the trail five days a week). The greenways have also become, I learned, a vital way to connect small communities.

Our first stop along the particular 21 mile rail-to-trail we are visiting is the town of New Freedom. It once was a vital hub along the railway. But trains no longer run. What's left of the town's former life is a caboose that sits like a flashing red memory on the side of the tracks. But today, the rail-to-trail is re-knitting communities along the train's former path and even helping to reinvigorate the town's history. The rail station never would have been refurbished had it not been for the railway.

I first meet Jerry Herbert. He'd opened Sara's Country Store just four months earlier, but the airy barnlike space is already packed with products and customers. It's the first grocery store the town has had in ten years. As Jerry and I talk, customers keep coming up to the long deli counter flanked by bulging bins of pickles. Jerry greets each one of them by name. "Wal-Mart doesn't know your name," he says to me later.

"And if we weren't meant to know each other's names—why would we have them?"

Down the block, Ed Hughes and his wife Kay introduce us to the Whistle Stop Bike Shop. They opened it a decade earlier after Ed, a trained electrician, lost his job. Unemployed in a fragile economy, Ed decided to start his own business. With a passion for biking but little business experience, Ed started the Whistle Stop. Today, the store is thriving thanks to Ed and Kay's dedication to their customers and their prime location: the store sits right alongside the trail.

"I understand that some families can't always afford our prices," Kay tells me, "but I ask my customers: if you have to buy your helmet at Wal-Mart, at least come in here and let me adjust it for you." Her customers are like her family, she says.

As we pull out of New Freedom the next day, I'm brimming with the sense of possibility: with a vision of the multiplier effect of building strong links locally. Sara's Country Store, the Whistle Stop Bike Shop, LaMotte's family restaurant across the street and the bed and breakfast inn around the bend: they are all examples of making this happen.

In the backseat of our minivan, I trace my finger along a map dense with roads; we seem far away from any thoroughway. I picture us deep in rural Pennsylvania. Then I lift my head and I see them: the tell-tale golden arches, the oversized Wendy's burger, and just beyond the fast food joints and gas stations, the aptly named GIANT supermarket. Below the crest of another hill I can just make out the sign for the football-sized Wal-Mart.

Suddenly, Kay and Ed and Jerry's Wal-Mart comments make sense: their stores are five minutes away from the nearest one. I had no idea how close we had been.

What does this story have to do with micro eco-farming? Everything.

Like the local economies flourishing along the rail-to-trail I visited, Barb brings home for us another emergence of "local": the flowering of small, vibrant eco-farms around the world—with a focus on the United States and Canada. These farms are creating something akin to what

Kay, Ed, and Jerry have: a renewed connection to community, to the human scale, to the earth.

And like the Wal-Mart just around the bend from New Freedom, the giant agribusiness farms are breathing down the necks of small farmers everywhere, and yet, as this book shows, these "new small farms" flourish.

So as we celebrate this emergence, we can also be clear-eyed about the challenges. We can remember the real threats to sustainable farming in this country. We can remember that we had so few farmers left in the United States that the census bureau no longer bothers to account for it as a separate profession. Big farm proponents tell us they're more "efficient," but we can remind ourselves that the real innefficiencies of centralized mega-agribusiness are hidden. They are not paying the environmental costs—to air, water, or soil—that their farms wreak. Nor do they pay the human costs of lost livelihoods, workplace accidents in processing plants, or decimated rural communities.

"Big" is easy to see; the "small" is more hidden. Small is the organic kitchen gardens in the back of homes that grow food for a whole family.

Small are the green roofs sprouting up on rooftops throughout the urban world—high, hidden and abundant. Small is a community garden plot that brings an immigrant family foods from home—the herbs and plants that are a vital part of their culture. Small is the farm down the road you've never noticed before that is making an artisanal cheese, almost all but forgotten.

Small in an age of big was an endangered species (but has there been a turnaround?). Barb reveals the small. She makes the invisible visible. Without books like hers to show us that small is alive and flourishing, it would continue to be unseen by most of us. Visible, we can celebrate it, join it, and support it. With the dozens of farms Barb describes here and with the stories from New Freedom in my mind's eye, I feel ever more empowered to do so.

—Anna Lappe, co-author of *Hope's Edge: The Next Diet for a Small Planet*

Introduction

Scanning the farm, I saw no reason why today would somehow be different; yet a curious feeling persisted. Trillions of dewdrops sparkled on beds of ruffled, rose-tinged Lolla Rossa lettuce heads and lacy green carrot tops. I operated a part-time market garden on about half an acre. This morning as the sun baked into the soil, the herb garden's fragrance mixed with the earth's scent and rose in steam around me. Birdsong mingled with honks from the white China geese who rustled on the straw nearby. Jeff, the only gander, played an important role on the farm—but not the role one might think. I didn't raise goslings, so there was no need for a breeding gander. Nor would Jeff ever serve as Sunday dinner. He was kept around because he was just too good at letting me ask him the meaning of life. Jeff had a way of making me laugh, of mocking the altered ego, and therefore reminding me not to find the answers there. Now, I asked Jeff what was different about today. He offered his usual answer of strutting as though he owned the world. This always reminded me that the answers were not in owning the world, but in owning the grace within oneself instead. So I turned within. A message in one form or another was trying to surface. What did it want to tell me?

Cedar and alder surrounded the green meadow I call Island Meadow Farm. As a "microfarmer," I had found a niche to serve. And as a professional writer, the outdoor work helped incubate my mind as I contemplated what I would write about next.

The author's son, Jereme, raised Jeff from a gosling.

On Mondays, I sent boxes of fresh-picked gourmet and heirloom vegetables by overnight express to subscribers in Seattle, about 90 miles away. On Wednesdays, local subscribers stopped off at the farm to pick up their bushel basket of produce I sat out in my horse-drawn wagon, giving them a bit of pre-industrial nostalgia for coming to the farm. On Fridays, which was today, I picked herbs, greens, baby squash, heirloom tomatoes, French market carrots and edible flowers for the weekend crowd at our local tourist town restaurants. My farm was an eccentric quirk hidden from corporate agribusiness. It was invisible to them—that's why they didn't take me seriously. That's why they left me alone to play. I was happy. But today, something was unsettled. It seemed a knowing was trying to push its way to me.

Before the birth of my farm, I had researched what, in this world of million-acre, machine-operated farms, a person could do to earn income if she remembered Grandpa's farm. There, her own hands would accept direct from the earth the fruits of the seeds she had planted. She would conduct a multi-faceted system already composed by nature, and each morning bring new vine-ripened jewels in reds, yellows, pinks, greens, purples and oranges to offer fellow humans in exchange for them supporting her craft of love. To find such a possibility, I looked for the holes left amidst the boulders of corporate agribusiness, and was delighted to find niches and gaps everywhere. I collected descriptions of a number of small quirky farms from which I drew inspiration to weave my own little farm. And then, also being a writer, I eventually published the descriptions in a pamphlet entitled, *Ideas for Small Farms.*

I wrote about an herb farm that allowed visitors to picnic there, a miniature horse farm that put on children's programs, and an apple orchard that thrived by growing antique apples and inviting visitors to a farm festival. I told of urban lot farms and backyard shed farms. I wrote about my own farm, even about Jeff. I described small, thriving farms that could have been locally adapted and duplicated in every region across the country. The list got up to 60, when I finally made myself stop. Where were they all coming from? Why hadn't I seen them before? I realized they, too, were graced with a sort of cloak of invisibility—invisible, that is, to the general mainstream, where there linger those who might see them as a threat. Here, in the safety of their invisibility, they incubated, they improved, and they grew in numbers. Collecting the examples was like playing in the sand that filled the gaps amidst the boulders. There was an endless supply of shapes and forms one could build into castles, unlike the unmoving corporate boulders. The potential list was endless. So I typed up the descriptions of 60, and the pamphlet was listed in *Books In Print,* while three copies were sent to publications that cater to gardeners and innovative farmers.

That's when something strange happened.

And the thought of that strangeness felt somehow related to the message trying to reach me. I had stopped at 60. But how many other little separate, eccentric farm quirks were out there that knew nothing of the others' existence? Did they need to know? Did those small-acreage farms that were making a comeback, those urban minifarms and backyard garden businesses, know there were others like them, and other people who wanted to join them? Did it matter... yes, the message was hinting louder. It mattered very much. I didn't know WHY it mattered so much yet, but from I-don't-know-where, orders for this humble pamphlet started coming from everywhere. Bookstores special ordered; newsletters featured it; universities requested it, and orders even came from overseas. It seemed there were not only more of these little guys around than I had originally thought; there were more guys and gals wanting to start little farms.

Out in my field, my harvest baskets were now mounded with herbs, flowers, fruits and vegetables. I cut one last huge bronze and green sunburst of "Sierra" lettuce, a Batavian gourmet variety that always delighted my customers, and carried the baskets to the house to rinse off and sort. That's what we microfarmers do; we grow things that delight and serve the people who will look us directly in the eyes. Some of us were part-time farmers, some even close to hobbyists. And some were operating full time, such as those who would show up at tomorrow's farmers' market held each Saturday in my small island town. Booth after booth would be loaded with produce. The round baby carrots there were so lusciously brittle they had to be hand dug, and so full of the earth's sugar that children ate them like candy. As the season went on, the children at the farmer's market would also taste the life force within Easter egg radishes and brilliant globes of vine-ripened tomatoes, and just-picked sweet corn with plump kernels waiting for their teeth to burst them open. Their parents would talk over recipes and discuss new crops with the farmers, who would in turn go home with a wealth of market research on what humanity wanted them to grow next.

We don't need corporate agribusiness to save us from starvation. Food is our excuse to co-create with nature instead of being passive recipients; to reach across species; to mingle with other humans; and to listen to an earthly rhythm.

Why on earth, at the thought of this beautiful scene, did anger surface in me?

"Is this what corporate agribusiness saved us from?" I asked myself (since Jeff was nowhere nearby).

We've come to think that entity will save us from starving. But it isn't lack of food we need to be saved from. Long before corporate agribusiness, the Incans fed 15 million with a three to seven-year surplus, and the Chinese fed one billion on less than 10 percent of their land base. We already knew how to feed and nurture ourselves from very small-scale, locally based production. And now, with new ways of networking and swapping information across the world, we know how to do it even better than the Incans. Charles Wilber of Crane Hill, Alabama, organically hand-tended his garden to harvest averages of 342 pounds of tomatoes per plant as compared to the California Tomato Growers Association estimate of eight to ten pounds per plant from commercial growers. Charles' advice on how to accomplish this is to look closely at nature and "pay attention."

No, we don't need corporate agribusiness to save us from starvation. Food is our excuse to co-create with nature instead of being passive recipients; to reach across species; to mingle with other humans; and to listen to an earthly rhythm. Now, I simply had to find Jeff and ask him why something so disconnected to that rhythm took over in the first place, and this time he gave me the true answer. His reply was loud, aggressive, and too narrow-minded to back off and see the bigger picture. Perfect. Humans at one time gave in to something loud, aggressive, and too narrow-minded to back off and see the bigger picture. This anger was becoming passion. The message was getting closer.

Now, just like Jeff, it wasn't that the traits of large-scale commercial agribusiness served no good purpose. Jeff and commercial agribusiness just got confused sometimes as to how much of the planet they owned. Corporate agribusiness could have created a supply of a few staple foods, great in cases of occasional emergency and temporary fear of starvation, and could certainly co-mingle among a foundation of strong local economies. But when we became dependent on corporate agribusiness, we said good-bye to the planet who was once evolving along with us.

So, separation from this Goddess Earth needed mending. The message coming was getting louder, but still not completely clear. How would this mending take place?

Now inside my farmhouse, thoughts incubated as I lifted sparkling heads of lettuce from the washtub. Empty wicker baskets in the shape of round over-sized trays lay like blank canvases as I instinctively arranged each garden offering artistically, alternating the greens with red, and tucking the brilliant edible flower petals in the center. I could hear the crispness as each head was set in its place. The pleasures of being one's own boss are many, and I thought of the home business renaissance spreading across the country. I realized I was engaged in an original home business, and I felt an ancient empowerment as the last head of lettuce was set into place. Being an eccentric quirk always made me feel I was running away from home. Had I been returning home instead? Or is the feeling more like getting back on track towards home, towards a place everyone is headed for in so many different ways?

There was a knock at the door. My neighbor entered and sat cross-legged on the floor as I rinsed the carrots and offered her one. Her gaze remained fixed on the abundance I'd soon deliver to the restaurants. She chatted about the new barn going up at the neighbor's acreage. Told how her dog was getting along with the kids. Asked whether Jeff had chased any more salesmen out of the yard. But, strangely, her eyes didn't go off those overflowing baskets. She couldn't seem to take them off if she tried. I realized something. As she gazed, she wasn't seeing the product of an eccentric quirk, of something strange and brand new and up-until-now-invisible. I know the look of quirk. Like the time Jeff inquired about a wind-blown, plastic grocery bag, then found the handles wrapped around his neck. Thinking he could outrun it as it billowed in the wind behind him, the joggers passing by slowed their usually unbreakable pace with an expression of not knowing whether to stop and take in more of this, or to run faster than ever before and never return. That was the look of quirk. But it was not in this neighbor. Not now. She was seeing something else. What was it? That question would not be answered until my delivery.

On most days, my farming partner took the produce to the restaurants. But today I would fill in, and one particular restaurant had us take our baskets directly through the front door to the chef. That's where the answer to my neighbor's questionable fixed gaze would be found. The morning restaurant patrons were engaged over coffee and rolls as I backed into the door with arms full of bounty. There were business-

I realized it wasn't just those of us who love tending the earth who were once again hearing its rhythm and paying attention... It was the unified human soul calling itself to reunite with the earth.

women, businessmen, nearby shopkeepers, accountants taking a quick morning break. I entered. Every head in the place turned and stayed fixed. I checked nervously to see what I may have dropped, or that I was dressed properly (or even at all). But again, here was that same fixed gaze at the garden abundance mounded into my baskets. It wasn't the limp, boxed factory produce snuck through the back door like most restaurant deliveries. It was the farmer bringing in her own creations that were crisp, full of shape and color, still illuminated with the sun's radiance. And again, their gazes were not of looking at novelty, but of looking at something else... They were, I realized at last, on the other side of seeing something novel. They were gazing with the look of recognition. Of remembering something lost. And of wanting it again.

Yet there was even more to it than that. When concert-goers gaze towards an orchestra, their eyes might glance first at an individual cellist, and then at a gleaming saxophone, but even as they zero in on one component, they are taking in the whole symphony. And that is how these people now gazed at me. They were not seeing just one farmer; they were taking in something much, much larger. And I realized it wasn't just those of us who loved tending the earth who were once again hearing its rhythm and paying attention. It was also the neighbors, the businesswomen, the businessmen, the shopkeepers and the accountants asking for direct contact with the farmer and land again. It was the unified human soul calling itself to reunite with the earth. This unified human soul was creating a new symphony, and that's what they were taking in when they gazed at just one farmer's baskets of bounty.

The symphony included the facts that local gardens on live soil were restoring oxygen to the air, undoing the greenhouse effect, returning the butterflies, restoring the pristine water supply, rebuilding local community, and returning a sense that something larger than ourselves was on our side.

The message's meaning was growing in intensity. While all of us little farming quirks were independently working our small, nearly unseen farms, we had been growing in number, and were in fact reinventing agriculture. We were reinventing the face of the earth, and the human village that flourished on it. Fueled by desires of human souls everywhere, the small, local, sustainable farm was being reborn. Never, once, had I been an eccentric quirk. I had been an incubation of the farm of a new century. All the others and I, though unaware of our connection, and beautifully different, were part of a huge new tapestry being woven across the planet. Like individual musicians in an orchestra, our differentness was not making us weaker, or less, but rather was creating a new melody that soared above corporate monotone, and that no one farmer could have created on his or her own, but contributed in a way that no other could.

The message now shot through as I remembered the following: Gerald Celente, founder of the Trends Institute, which correctly predicted the 1987 stock market crash and the break-up of the Soviet Union years before those events, said it well. He reported that by the 21st Century, organic microfarms of ten acres or less would begin to challenge the food giants. According to the USDA, farmers' markets, one of the best outlets for small organic farms, grew in record numbers and increased by 63 percent as the last century turned. That increase grew to 79 percent as the new century marched on. The Rodale Institute reported that community supported agriculture, a form of small, sustainable farming, was booming across America while the Bio-Dynamic Farming and Gardening Association reported up to 75 calls a day from citizens looking to connect with a small farmer near them. There were just so many gaps that corporate agribusiness could not fill, the small, local, personalized farm's time had simply come again in a new and better way. They were blossoming across the planet. As long as there were different bioregions and communities with different personalities, there was an endless supply of small farm possibilities, and endless reasons for both the farmer and the citizens they connected with to hear the earth again. The sustainable small farms, under the brilliant disguise

"Pile it high and kiss it goodbye:" Farmers' markets provide a diverse, friendly marketing outlet for lots of micro eco-farmers.

of quirky little niches that kept them invisible to those who once could have swallowed them up, were now everywhere.

The invisibility was there for our protection. We needed it in order to create ourselves. And that's what the message was that was trying to push through: *WE HAD BEEN CREATED.* The cloak of invisibility had been lifted. We had been recognized. We were being watched.

And now that we were a solid new entity, a new gift had to be given to us to continue the earth's evolution we were serving. Invisibility would no longer serve us as it once did. We now had to see each other, come together to create a new whole much, much stronger and larger than the sum of its parts.

There would now be a drawing together of those who tended and raised the organic goat dairy products from kindly raised animals, the historic nursery plants on one acre, the backyard flower patch, the U-pick citrus grove with free range hens running beneath it, the bio-dynamic vineyard with bluebird boxes for pest control, the rare-breed dairy cows, the hand-tended chocolate mint, the blue-fleshed potatoes, heritage roses, Jeff. We were so different that until now, we didn't know we were all listening to the same conductor, where our new strength, instead of invisibility, would now come.

After collecting cash from the chef, I returned home, and the next morning scanned the dewdrops, herbal fragrances and birdsong. But now I knew that each time I transplanted a seedling of "Sierra" to harvest one day, or added one more French market carrot to my basket, I was adding to the great symphony of the earth's upward evolution. As I created with the freedom and risk of an independent farmer, I could draw inner strength from the silent conductor that wove me to all the others. And I knew what I would be writing about next.

CHAPTER ONE

The New Micro Eco-Farmers

I love co-creating with the earth, as it is so simple and healing to live close to the mother earth that sustains and nurtures me every day with her beauty. I love to hold her in my hands and watch the seeds grow into fullness just by tending to them slowly along with the sun and rain.

—Mariam Massaro, Singing Brook Farm,Worthington, Massachusetts.

On less than an acre, Mariam Massaro tends certified organic herb, vegetable and flower gardens, which include more than 78 varieties of roses. Along with this, she raises Icelandic horses, llamas, Angora rabbits and Icelandic sheep in the farm's Berkshire Hills setting of western Massachusetts. The animals provide offspring and specialty wool for Mariam to create fiber crafts for sale. They fertilize and mow the gardens. In her 1850s New England farmhouse, a workshop overlooks a year-round brook. The herbs, flowers and wool are processed into products sold both locally and worldwide.

On five acres surrounded by woods, Sylvia and Walter Ehrhardt of Knoxville, Maryland created the successful Ehrhardt Organic Farm. From its earlist years, chefs could not get enough of their organic dessert quality blackberries, which bear up to nine weeks in their location. Chefs also gladly paid premium prices for their fresh-picked, locally- and sustainably-grown raspberries, strawberries, miniature squash, carrots, tomatoes, herbs and shallots. Thousands of plants were started

The Chile Man is a ten-acre farm near Round Hill, Virginia that produces natural marinades, salsas, mustards, barbecue sauces, and pesto from peppers, berries, and herbs grown on its sustainable farm. Products are manufactured in an on-farm, FDA-inspected commercial kitchen and are sold at festivals, online, at farm events and in specialty food stores.

Several years ago Robert Farr, 45, with his wife Carol and two children left a career in high-technology marketing to pursue a new life and a dream of self-reliance on the land. Check out the farm and Robert's practical tips for "Making It On Ten Acres" at www.thechileman.com.

each spring in the greenhouse attached to their living room, just a short walk from their growing area. They grossed $12,000 an acre on miniature squash alone. Their blackberries produced two tons per acre per season. The farm, still going strong, became a demonstration farm as well, and their reputation as successful organic growers spread nationally and internationally. "Over the years, we found that we didn't need to expand our land but to make it more productive," said Sylvia.

Ocean Sky Farm, owned by Art Biggert and Suzy Cook of Washington State, is a 1.55-acre suburban full-time farm. They have operated a highly successful community supported agriculture (CSA) farm where 75 families "subscribe" to the farm by paying upfront for weekly delivery. Being microfarmers, they can adapt easily to new interests, and eventually chose to incorporate other farm products including perennial medicinal herbs. "People come to the farm, take what they need, fill out their own receipts and leave cash or checks in the tea pot," Art said.

> *There is a change among those who farm in this century. They are not different than what they used to be. They are more of what they always were... they seem to have taken a long-lost power back.*

Baruch Bashan, creator of Gaia Growers Farm in Portland, Oregon, had been a part-time vegetable gardener for more than 25 years. One year, he produced 2,000 pounds of vegetables on two city lots. Not needing it all himself, he ended up donating it to the local food bank. "That last year I decided I'd had enough of working in an office as a software programmer, and wanted to do farming full time," he said. It wasn't until July of that year that he secured a half acre, but still had a successful growing season and launched Gaia Growers Farm. He was convinced, as he stated, that "a single, hard-working person can run a successful organic veggie and seed growing business on small acreage, without having to invest a lot of money."

There is a change among those who farm in this century. They are not different than what they used to be. They are more of what they always were. Some come from generations of farmers or gardeners. Some have just joined those who earn a living from the earth. Regardless, they seem to have taken a long-lost power back. Are you one of them?

"After 18 years working in the corporate world, I'd had enough," said Robert Farr, also called "The Chile Man." He owns a ten-acre sustainable farm in Virginia, and here he shares his story with us as he also wrote it for *American Farmland Trust Magazine.* "I'd always had a relationship with the land, hiking through the Appalachians and the Rockies, and I'd grown up with my hands in the dirt. But faced with the prospect of endless days in windowless cubicles, I decided to live a dream I'd had since adolescence, and start my own sustainable farm. As soon as we closed on the farm (July 1998), in Loudoun County,

Virginia, I quit my job as a marketing manager in the computer industry and The Chile Man was born." He now grows 67 varieties of peppers and other fruits and vegetables to produce more than 40,000 bottles of all-natural marinades, barbecue sauces, mustards and salsas right on his ten-acre farm.

The term I use for this book, "micro eco-farm," sprung from this change in farmers. From urban lots to small town backyards to rural small acreage, this term is the umbrella for highly abundant, constantly improving, ecologically operated microfarms that produce a mix of fruits, vegetables, herbs, grains, nuts, mushrooms, flowers, fibers, craft materials, organic, pasture-fed dairy products, farm-crafted creations, and farming education and experiences. The examples in this book emphasize farms from fractions of an acre to five acres that earn full-time income for at least one adult. Some provide the entire income for single adults with several children, and some provide the main income for two adults and their families. It also touches on microfarms that integrate with complementary home and cottage businesses, those that prosper on six to 15 acres, and a few who earn a supplemental income.

All are sustainable in a variety of ways and are taking traditional organic production to new levels. Whether they reintroduce ancient royal gardening techniques or are the first to profit from the latest U.S. research, they connect sustainable local minifarming with the care of ecosystems and entire world populations. Some even say it's as though they are recreating an advanced form of Eden. They are willing to work harder short term in order to have more time long term for further creativity and loving their friends, neighbors and family. This seems to be just what the earth wanted, anyway: A co-creation of human innovation combined with the earth's superior ability to "do the hard work." Even organic pest spraying, rototilling and weeding will become less and less necessary at the hands of these farmers.

Fueling this new entity—the micro eco-farm—are several supporting changes in human values. These include the environmental and health movement, the delicious "Slow Food Movement" (see Resources and Networking), the push to strengthen local economies and the

Salt Creek Farm is a small certified organic family farm located on the west side of Port Angeles, Washington. CSA program clients receive a seasonal abundance of fresh vegetables, herbs and flowers direct from the farm each week. Website: www.saltcreekfarm.org.

Photo: Salt Creek Farm owner Doug Hendrickson, taken by CSA subscriber Martin Hutten.

parents wanting their kids to connect to nature and their food source again.

We now know that large amounts of farm crops can be produced intensively on very small amounts of land very easily and very simply, and as this happens, the land and crops get more abundant year after year. There are many techniques that allow for this, and yet, micro eco-farmers don't always use just one of them. Often, they will synergize several, to create a new whole much more prolific than the sum of its parts.

Micro eco-farmers do not compete with mass-produced, underpriced products. As one would guess, they supply the niche markets. However, you will soon see that there are more niches than anyone ever dreamed of. These "tons of niches" collectively add up to a very large opportunity for new micro eco-farmers, almost making non-niche farms seem like the oddball.

These micro eco-farms, along with their larger sustainable agriculture cousins and sustainable home gardeners, choose the rhythm of a new drummer—that of the earth as the solution, rather than the earth as the problem. They still touch the soil; they still plant the seeds; and they still nurture the animals. But, because of the retrieval of their power, they have switched direction, crossing the bridge back home, rather than crossing the bridge far away. No longer running from the earth as one would run from an enemy, forcing and succumbing, they are now moving towards the earth as a source of latent and untapped wisdom.

Whenever they need an answer, the answer seems to appear—such as it did concerning the honeybee problem that began in the late 20th century. Honeybee populations were dwindling. The bees could not, it seemed, sustain their health and numbers, succumbing to parasites and other invasions. About this time, Adaline Harms had secured her five and a half acres on the edge of Mt. Shasta, in California. Here, she now gardens in her greenhouse and hexagon-shaped raised beds. Adaline is one of the most spiritual and earth-loving people I have ever met. My conversations with her remind me that whatever created this earth speaks to us in many ways, including directly through the earth itself, even through its own honeybees, if need be…

"I took a trip to Arizona, and while driving on Highway 5 the length of California," she said, "I kept seeing all the bee hives on the side of the road. I just got this feeling that I needed to keep bees. I knew absolutely nothing about beekeeping, so when I got home, I started asking around about beekeepers to learn from." This eventually led her to someone who had worked with Ron Breland, who has a nursery and bee sanctuary in New York State and who had developed an alternative hive. Ron reportedly noticed that in nature, bees don't build hives in the shape of file cabinets. So Ron mimicked nature's design in his hives, and his bees thrived well.

"How quaint," I thought before actually seeing this hive. I imagined something simple. Maybe something Winnie the Pooh would climb up and get his nose stuck into.

"So, Adaline, is it round, hexagon?" I asked.

"It's a dodecahedron," explained Adaline.

"A what?"

"A chalice made up of pentagons, with a similar shape turned upside down on top of the chalice to make up the brood chamber, then there are five-sided extensions that stack on top." Adaline had a hive built according to Ron's design and observed her bees gaining strength.

Like Adalene, these new sustainable farmers and gardeners are freer to be innovative again. Without thousands invested in equipment specialized for one specific crop, or fees paid to support large advertising firms that push a crop they are entrenched in, they can change crops, and they can change "equipment," on a summer weekend. The following year, Adaline's carpenter built four more hives.

"We've made a couple of minor alterations to the original design, while trying to stay with the original ideas and intention."

The differences among individual micro eco-farms are many, yet this is their strength. If you are about to become one, you will create something like no other. You may develop a purely vegan farm, supplying those who consume only plant foods with aromas, textures, proteins, micro-nutrients and "life force" in a variety previously unheard of.

Perhaps you'll operate a "Paleolithic farm," concentrating on nuts, roots, wild greens and other foods humans once consumed before grains became a mainstay. Maybe you'll add wild-grazed fermented dairy products as our ancestors once consumed.

In a world dominated by an oversupply of questionable grains, you may even grow grains. But yours may be pre-industrial grains such as spelt or quinoa, grown organically and intensively for higher production on smaller parcels of land, with the grain stone-ground right after harvest, right on the farm. In fact, you may even sprout your grain before it becomes bread, turning it back into a "vegetable" rather than a grain, to bake into loaves in your farm's own hand-built brick ovens.

Yours may be one of the only farms reviving food of the Incas, such as ahipa—*pachyrhizus ahipa*—fabales, a legume grown for its sweet, apple crisp roots, or arracacha—*arracacia xanthorhizza*—apiaceae, which looks similar to celery with uniquely flavored roots, or maca—*lepidium meyenii*—brassicaceae, with tangy, radish-like roots.

Or maybe you'll operate a farm that provides for Italian cuisine chefs, gourmet hobbyists, or local and online ethnic groups. Your unique herbs and vegetables can't be found in supermarkets. You'll provide those vine-ripened Italian tomatoes even in winter in your 10 x 10 ft. greenhouse, while teaming up with a neighbor whose goats fertilize your gardens, and who creates boutique cheeses that complement your Italian sauces sold throughout the winter months.

You may produce products for other farms and gardens: worm castings, heirloom vegetable seedlings, locally-adapted garden flowers, and heirloom seeds, to name a few. Some farms provide "experiences" even more than products, with a children's pony farm or an herb farm with herb related classes. The selling of experiences works well for those microfarms that attach to larger established businesses such as destination spas, schools, campgrounds or spiritual retreats, that automatically draw in visitors as part of the farm income. However, "microfarms within larger establishments" can also be, well, "microfarms within larger farms."

Zestful Gardens, located near Tacoma, Washington, is a diversified small scale farm specializing in annual vegetables. According to owner Holly Foster, the farm uses Biodynamic techniques and intensive cover cropping; produce is marketed through a CSA, farmers markets, and restaurants. Contact: hollykfoster@hotmail.com.

Theresa and Matthew Freund own a Connecticut dairy farm. When they filled a wagon with their garden's extra-sweet corn on the side of the road, customers stopped to buy, and also asked for lettuce, tomatoes and cucumbers. Following this lead, they planted more of the things they were asked for. Eventually, their farm stand took in $100,000 over the summer, while Matthew and his brother continued to operate the dairy. The Freunds expanded their roadside stand into a two-story barn-type building offering their fresh produce, dairy products, jams, vinegars, and they even added a U-pick wedding flower acre.

You may be a microfarmer who does not even grow food. Some produce ornamental wheat instead, or herbal goat milk soap, naturally-colored cotton or Angora wool.

"In the US, there is a lady who 'paints' pictures using flower petals out of her garden. Another farmer grows seven acres of broomcorn, makes brooms and sells them retail and wholesale," said Ken Hargesheimer, who teaches sustainable mini-farming, mini-ranching and market gardening in the US and worldwide.

"There is a grassroots movement back to family farming," he continued. He has seen for himself what he describes as "free enterprise and micro-entrepreneurship" in both urban and rural environments where mini-agriculture has been proven to produce substantial income on surprisingly small parcels.

"People can have a comfortable income, a high-quality lifestyle, and a great way to raise children," he said. "As well, the micro eco-farm can adapt to year-round work, second-family income, spare-time income, or even full-time income for part-time work." A lady took a (mini-farming) course," Ken said, "returned to Alaska, prepared her land and grossed $20,000 the first year, and then had a six-month winter vacation!"

Micro eco-farms team up nicely with other cottage industries. Personal chefs can create one-of-a-kind cuisine from their own mini-farms. Massage therapists can create their own line of garden-fresh, body-care products. Bed and breakfast inns are very popular when combined with small working farms, each enhancing the customer draw and promotion of the other.

Micro eco-farms fill in spaces that larger sized farms don't attend to. They use back yards, vacant lots, or their family's own small acreage. As they grow in number, it is anyone's guess as to what type of new economical foundation they could create.

Regardless of their differences, they have one thing in common. They seem to have an inner knowing that creating with the earth is attached to Something Greater than anything purely human-created, and they must continue to work with this greatness. As they do this, their presence on the earth is collectively creating a very beautiful world.

"In the past four-and-a-half years, we've seen our holistic farming practices dramatically increase the bird population," continued Robert Farr, "One of the most rewarding parts of being a small farmer is the opportunity to be in constant contact with the spiritual; to do, as Gary Snyder best said, 'the real work.' I need only stroll out my kitchen door to be immersed in the holiness of nature, to see the mountains, endlessly walking. All of our own sacred nature begins outside, in the worship, as the Amish say, of God's creation."

It often doesn't feel so much like a business separate from leisure and hobby time, and it gets less and less important to distinguish "work" from "play."

"It's a relationship," said Diana Pepper of her 2.75-acre Green Frog Farm in northwest Washington State. Diana reminds me of a "human faerie" and is a living library of earth wisdom. While she and her partner, John Robinson, occasionally wild craft their acreage's native woodlands and meadows, most of their production is on only one-third of this acreage. Diana and John have established a Pacific Northwest native plant nursery, selling native trees, shrubs and groundcovers, plus herbs and ornamental flowers. They also create small bottles of herbal and flower healing products, kits for massage therapists, offer workshops, and private consultation sessions. Both agree they are not separate from their livelihood: "It's 110 percent of who we are," said John.

At this point, micro eco-farms fill in spaces that larger sized farms don't attend to. They use back yards, vacant lots, or their family's own small acreage. As they grow in number, it is anyone's guess as to what type of new economical foundation they could create. We are currently still dependent on a system that produces a few staple crops on huge acreages that ship these crops across the country and world. Jo Robinson, author and educator, states it well, "We need micro eco-farms, midi eco-farms and maxi eco-farms."

The current problem with food production is not that there isn't enough food, but more that it isn't produced where it's consumed fresh off the vine by the region's own local citizens who are in tune with that Something Greater, making the local growing decisions, choosing the locally-needed adaptations, and keeping the food and revenue close at hand. When a stable local economy is created this way, distant shipping to and from far away lands becomes a friendly trade rather than dependency. The ability to produce locally is one of the many gifts of the micro eco-farm. And with world travel and technology that allows networking among eco-farmers of all sizes, their successes are mounting at an accelerated rate.

It is my honor to present in the following chapters a close-up of those who are actually succeeding, a treasury of what you can grow, what farming methods you can use, what animals you might like to choose from, and how the farmers reach their markets. I will present an emerging new foundational how-to on growing methods that span all forms of sustainable small farming, no matter what is grown or where the location. Then I will distill many methods that have increased production on small ground from double to up to 40 times that of conventional growing. You can choose which ones you want to explore and incorporate. Mix and match, and see if you can make two plus two equal 10, something you will see demonstrated in Chapter 6. Once you see all that is available to you, as with every farmer I interviewed, you may find that what's inside you is the greatest success secret of all.

"I love to create. I'm strong-willed and muse-driven," said Baruch Bashan, of the above described Gaia Growers Farm. "I got into software, like one does writing or painting. And, as with those other artist-types, having some other person decide what you create ain't quite the same thing as when your Muse calls. So, this allows me to define what is to be created." It's as though these new micro eco-farmers sense something on the horizon that is beautiful, and they are taking us there.

CHAPTER TWO

A Treasury of New and Ancient Crops

We focus on heirloom varieties and are working to develop new lines of certified organic seed. Our heirlooms include "Country Gentleman" yellow corn, "Amish Salad" tomatoes, "Thai Pink" tomatoes, "Poona Keera" cucumber, "Charentais" melon," Lakota" winter squash, and "Blue Ballet" squash. Plus, culinary and medicinal herbs, sunflower and amaranth bouquets and PUMPKINS! We are growing everything from the "Wee Be Littles" to "Sugar Pie" pumpkins to the "Big Max" pumpkins.

—Shelly Elliott, Idle Thyme Farm, Carpenter, Wyoming.

On the newly created Idle Thyme Farm on the prairies near Carpenter, Wisconsin, Shelly and Kevin Elliot have plugged into a huge window of opportunity in the world of plant crops for this century. "We are planting 2 – 2-1/2 acres this year," said Shelly. Their crops include organic, fresh herbs and vegetables, salad greens, potted vegetable and herb seedlings, and a "harvest time" pumpkin patch. And this is only a hint of the possibilities available.

The Return to local production, where the window first begins to open

Corporate agribusiness grows crops bred for tough mechanical harvest, uniform processing size, and long-distance shipment. Micro eco-farmers' crops do not need to endure rough, long-distance shipment. Nor do they need to comply to uniform shapes and harvest times. Even when they create on-farm products from their crops to mail out, or to send carefully packed, fresh shipments overnight, their custom packaging does not require tough crops nor uniform sizes. Therefore, local micro eco-farms can offer tender garden crops rather than corporate agribusiness crops. This is where their opportunity begins. They hand harvest or gently harvest their crops and sell direct or to local buyers immediately after harvest. So not only are their crops more richly flavored and colored from mineral-balanced, live fertile soils and further enhanced by being vine- and sun-ripened, and not only do they have their personality and unique farm expression attached to every crop they sell, but the crop choices they offer their customers are astounding.

Heirlooms

Between about 1870 and 1920, there was a golden age of gardening and farming in America. "Early Jersey Wakefield" cabbage came early and was so sweet, kids ate its leaves for snacks. "Lemon" cucumbers provided round, sweet, and mild fruits the size and shape of lemons. Hundreds of tomato varieties were adapted to every climate and came in colors of bright apricot orange, lemon and golden yellow, purple, pink, white, golden green, striped and reds of many sizes and shapes. Lettuce was crisp and tasted of life. It came in miniature sizes and giant sizes, frilled and smooth, rose speckled and deep burgundy tinged which accented the forest and lime green varieties. Apples came off the tree already spiced with apple pie undertones and aromatic flavor blends made into exquisite ciders. There were peach varieties that withstood spring freezes, allowing locals to taste neighborhood peaches picked at their peak, rather than green ones picked and shipped from miles away. There were peach varieties with such a meltingly delicious flesh, sweetness and aroma, that the current peaches bred for a clinging inner stone and hard texture for shipment almost seem like a different fruit. Hundreds and hundreds of varieties prevailed.

Overseas, the gardens and farms had produced their golden gardening ages, also. Mints and basils came in flavors of chocolate, lemon, lime, licorice and cinnamon. Radishes looked like multi-colored Easter eggs and round French market carrots were sweet, brittle and aromatic. Artichokes were tender and purple; Italian broccoli had bright green spiraling heads; and Italian cauliflower was mineral-rich and brilliant-purple. Chinese cucumbers grew to 19 inches with few seeds and superb flavor.

When corporate agribusiness took over the regional citizens' farms, producing most of humanity's food from a few centralized locations and then shipping them across the country, it is estimated that about 80 percent of those delicious cultivars eventually became extinct. Gardeners and farmers stopped saving their own seed and the independent seed companies that provided those varieties were purchased by larger seed corporations that dropped the garden varieties in favor of corporate agribusiness crops. The choices of crops to grow for both farmers and gardeners became less and less. Flavor, aroma, texture, beauty, health, tenderness, juiciness, local adaptability, thin melt-in-your-mouth skin... all succumbed to commercial hybrids bred for long-distance travel, mechanical harvesting and factory-uniform shape. Strawberries now had to withstand a voyage of thousands of miles from California to New York. Sweet, tender, garden and locally farmed varieties fell to the wayside. Carrots had to be tough enough for trucks to accidentally drive over. The baby market carrots, brittle and richly flavored, were forgotten. When these authentic foods and their flavors disappeared from American cuisine, chemically created laboratory foods tried to fill their spot, with brightly colored dyes and sugars and exciting artificial flavors to fill the emptiness, and unhealthy but commercial-friendly fats to make up for the natural craving for healthy essential fatty acids found in just-picked produce.

Yet new cultural values sparked a renaissance to return to a golden age of gardening and farming even better than before. Hidden in jars and envelopes, and growing in quiet generational gardens, the seeds of many of these lost gems were still available from private citizens, as well as new ones no one but these gardeners knew about at the time. Little by little, new heirloom seed companies sprang up again from the offerings of those private gardeners. They gathered these seeds and multiplied their supply. The heirloom fruit and vegetable variety numbers are growing again at a delightful pace.

On the Carey Family Farm in the Oregon Cascade foothills, lost varieties are a main focus. Like the Italian pimento: "It's great in salads, has a robust flavor without being too hot," said Nancy Carey. "I grew up on the Monterey Peninsula in California where the Italian population was as plentiful as the Hispanic. We learned to like a wide variety of ethnic foods."

The movement began near the end of the last century to hunt these varieties down, increase their numbers, and spread them across the world again. Gardeners, seed-saving clubs, micro and small farms, and small businesses emerged to fill this need. "We grow around 100 rare varieties each year on our farm," said Jeremiath Gettle, owner of Baker Creek Heirloom Seeds and publisher of *The Heirloom Gardener* magazine. "The rest we buy from more than 20 small farmers, small companies and gardeners, and have first-hand knowledge of nearly everything

Renee Shepherd is widely regarded as a pioneering innovator in introducing international vegetables, flowers and herbs to home gardeners and gourmet restaurants. The seed line offered at Renee's Garden is her personal selection of new and unusual seed choices of time-tested heirlooms, international hybrids and fine open-pollinated varieties. "I offer only the varieties that are very special for home gardeners," Rene says, "based on great flavor, easy culture and exceptional garden performance." Website: www.reneesgarden.com.

we sell." Jeremiath traveled to the Thailand countryside to bring back more than 30 varieties from Thai tribes, including a red-skinned Thai melon and unique eggplants with names like "White Egg" and "Green Striped Egg." Seed saving magazines and clubs teach how to collect local heirlooms from one's own neighborhood. "Market farmers and gardeners have more than 100,000 choices," Jeremiath said.

Jeremiath described to me a small farm he has seen first-hand that added heirloom squash to their pumpkin patch. They now raise 350 varieties. "Up to 100,000 people come each year from several states to see the most incredible display of squash, pumpkins and gourds," he said. "On a busy day, they can have five long checkout lines, selling heirloom squash like it is the latest fashion."

These crops are "open-pollinated," versus "hybrid," meaning their seeds can be planted the following season to produce more of the same, with the best ones selected to continue on with an improvement in performance and flavor. Open-pollinated crops that have been maintained for more than 50 years are considered "heirlooms."

"People are tired of styrofoam-like tomatoes that store like rocks and taste like chemicals," said Jeremiath. "When someone tries an heirloom, the first thing they notice is the flavor... wow! There is no comparison," he continued, noting that heirloom varieties often win out in taste tests. "They are also colorful. That's why sales of our heirloom seeds have been going up at a rate of over 60 percent a year!" He described to me a small farm he has seen first-hand that added heirloom squash to their pumpkin patch. They now raise 350 varieties. "Up to 100,000 people come each year from several states to see the most incredible display of squash, pumpkins and gourds," he said. "On a busy day, they can have five long checkout lines, selling heirloom squash like it is the latest fashion."

Tomatoes are a good example of how a currently mass-produced crop still has room for the micro eco-farm to produce that crop prosperously. They are considered the favorite backyard garden crop and are certainly plentiful at the supermarkets. There are plenty of chemically grown tomato hybrids or genetically engineered tomatoes from Mexico, Florida or California. They are selected for thick, hard, dry flesh for shipping across the continent, often picked green and gassed with ethylene to initiate ripening, or at one time reportedly crossed with fish genes to create the appearance of vine-ripened, freshly-picked fruit at the end of a long trip of many miles. Most nearby supermarkets carry these inexpensive, "fresh" tomatoes year-round. But Stuart Dickson, one of the first pioneers of the new micro eco-farm movement, grossed $70,000 from two acres of tomatoes on his Stonefree Farm back when supermarket tomatoes were plentiful, and minimum wages were around three and a half dollars an hour. His organic heirloom tomatoes are niches that outshine those mass produced.

A microfarm's tomatoes may include the big old-fashioned backyard beefsteaks, slicers, stuffers, early ripeners, or natural long storage tomatoes that ripen slowly indoors over the winter. The farm may offer heritage varieties in a rainbow of colors and collection of shapes including "Golden Queen," an orange slicer dating back to 1882, and pink "Zapotec," an ancient Native American variety.

As an example of another common fruit's potential, the melon, here's a sampler of what could be offered in your heirloom melon patch:

- "Green-Fleshed Pine-Apple:" a very rare heirloom grown by Thomas Jefferson and later sold in America and France in the mid-1800s. Highly perfumed with sweet firm flesh.
- "Valencia Winter Melon:" an heirloom believed to have come from Italy, which has very sweet cream-colored flesh and can keep four months into the winter.
- "Ginger Pride:" melons can reach 20 pounds, with yellow-orange skin and sweet melting flesh. It was grown by early American families who carefully selected and saved the best seeds of this mammoth variety.
- "Tigger:" this heirloom from an Armenian mountain valley market has brilliant yellow rinds with fire-red zigzag stripes. Its intoxicating fragrance can fill an entire room.

- "Orangeglo:" a deep orange-fleshed watermelon that is naturally resistant to insects and disease, with a tropical flavor and considered the best flavor of all by many.
- "Kiwano (Jelly Melon):" This one has transparent lime-green flesh and a sweet-sour mix of lime and banana tropical flavors. From Africa, it grows anywhere melons grow.
- "White Meated:" White fleshed watermelons grew wild in Africa and were sold in America in the 1800s. This melon has a unique fruity flavor with juicy white flesh.

There are many more melons—150 alone are depicted in Dr. Amy Goldman's book, *Melons for the Passionate Grower.*

Along with annual food crops, this return to richness also includes "heritage roses," antique flowers, and other landscape plants. Betty Adelman of Waterford, Wisconsin expanded her interest in plant histories into a small business she named Heritage Flower Farm, which specializes in plants that existed before 1900. Her collection of plants includes those favored by Roman emperors, Native Americans, and plants mentioned by historical personalities such as Thomas Jefferson and Emily Dickinson.

This renaissance is also restoring "antique" apples and other tree, cane, bush and vine fruits.

On my Island Meadow Farm, people are astounded to pick a "Sweet Sixteen" apple right off the tree and taste the undertones of cinnamon, cloves and nutmeg. Children adore biting into the several pink and red-fleshed apple varieties we grow.

"We grow the old fashioned Gravenstein... yellow background with red striping," said Nancy Carey. "It's a favorite because of its flavor in apple sauce and pies. My mom and dad always used them and my husband's mom made the world's best applesauce from them. The smell and taste bring back so many great memories."

While supermarket apples are cheaply mass-produced, a micro eco-farm might instead create apple butter made from favorite varieties grown on Monticello. Antique apples and gourmet apples provide customers with bright pink flesh, natural immunity to apple scab, apricot-colored flesh, and tropical-flavored undertones. There are apples that are superior for baking; those that ripen in June; those that ripen at the end of November; and apples that actually "ripen" in winter because they mellow and improve in flavor if picked at their peak and gently stored in the home cellar. When the 20th century system bred apples for shipability and uniform shape, out went the tangy-sweet, aromatic and juicy apples such as "Cox Orange Pippin," "Pound Sweet" and "Tompkins Country King." Tom Berry grows five acres of rare, specialty apples on his Canyon Park Orchard in western Washington State. The rare apples are often presold before harvest over the internet and customers pick them up right on the farm for two months during the season.

Culinary specialties from local communities, wild crops, seed-saving, and cuisines and cultures from around the world.

Heirlooms, antiques, vintage and heritage crops play a large role in the food crops micro eco-farms offer. As mentioned, home gardeners had quietly kept them alive once their continuation was dropped by large American seed and agriculture business. And along with heirlooms, there are places in the world where the craft of breeding brand new crops for flavor, beauty, tenderness, aroma, uniqueness and local adaptability has continued on as small, ethical businesses. This craft of breeding may still start with heirlooms; in addition, it often reaches out to discover, test, name and proliferate previously nameless garden varieties. The craft also breeds known varieties "backwards" to try to restore extinct traits, then tests, improves, and creates new crops with new names. Gourmet and connoisseur seed companies offer these new choices along with their favorite heirlooms. Quite often, their offerings are open-pollinated, meaning they, too, can be reproduced by farmers who may prefer not to become dependent on the company as their only seed source.

Along with this, farmers themselves are breeding their own varieties, selecting the best from open-pollinated crops and improving them year after year. "We focus on heirloom varieties and are working to develop new lines of organic seed," said Shelly Elliot of Idle Thyme Farm. As well, local wild crops are making a comeback. Microfarms may nurture native wild-trailing blackberries, wild mushrooms, and local sea buckthorn berries. There is an expanding list of fruits, herbs, flowers, vegetables, nuts, mushrooms and other flora products that cannot be produced on a large-scale, centrally located system.

All of this paints an ever-growing picture of a returning Eden. Forgotten native fruits are returning and crops that grow naturally in specific bioregions without compromising the local ecosystems are flourishing. Pineapple guava and kiwi fruit grow in the Pacific Northwest. Aromatic African melons and gourmet European white- and rose-colored asparagus grow on neighborhood American microfarms. If farmers prefer not to save seed, a growing number of independent regional seed companies now offer them again. Nature has inserted in plants the ability to change a touch each year according to the local bioregion, which is why independent seed companies that perpetuate crops for local adaptability are so valuable.

Occasionally, hybrids are also chosen by micro eco-farmers. Hybrids are a natural cross between two parent plants that cannot be counted on to reproduce that same plant the next year (most seed from hybrids results in one of the parent crops instead). Nature creates hybrids naturally during cross-pollination, and farmers can create their own hybrids. Each season, the two original parent plants are crossed to provide next year's seed, and the seed is collected. Eventually, after years of trials, some of those hybrid seeds may reproduce that new combination from seed. Then, those seeds are proliferated and a new open-pollinated crop is born. When micro eco-farmers choose hybrids from seed companies rather than their own operation, they must stay aware they will have to repurchase them from that company each year; so they need to make sure that particular crop is not their main source of income in case the company discontinues that crop's seed.

Naturally, all micro eco-farmers avoid pollution from genetically engineered crops. Altering the genes in this manner is not speeding up

Plant starts at Claymont Community Farm, Charlestown, West Virginia. The farm produces and markets "fresh, local and consciously-grown" vegetables, fruits, fresh-cut flowers and herbs, as well as bread and other baked goods, pesto and salsa. According to owner Bevin Coffee, the produce and bread are sold at nearby farm markets, as well as through a CSA (Community Supported Agriculture). Website: www.claymont.org.

natural selection as designed by universal law. Universal law, instead, creates life that harmonizes with the entire universe and has checks and balances enforced to make sure certain barriers are not crossed. Universal law does not allow a tomato to mate with a fish nor a human to mate with a potato to produce outlaw mutations that may have lost their enzymes to break down in nature and in our digestive systems. Nor does nature allow outlaw mutations that have lost their ability to reproduce from seed (traits which may spread to the natural world from cross pollination), or are causing the reported death and illness of those living things that consume them under the notion they are healthy, natural foods.

Nor are these genetically produced crops (as well as chemically-grown crops) saving us from starving. We are not starving for more of that type of food. We are starving for food to be produced in abundance where it is consumed, and for the citizens to return their innovation, artistry, uniqueness and fertility to the land we live on.

Further niches and on-farm products.

Micro eco-farmers usually choose crops from this rich rediscovery of old-time varieties and the new co-creations continually pouring forth from all continents. Further, they break these crops down into unique niches. In every plant-based typical farming category (food, fiber, decoration/craft material, dye), there are niches which fill in the cracks between the boulders of that category. As an example, herb growing has a multitude of niches: herbs grown only in the Pacific Northwest, historical herbs, cosmetic herbs, Italian herbs, healing herbs, spiritual herbs. There are niche ways to offer them: U-pick, potted live, weekly delivery, hand-made into tinctures and salsas. There are local niche markets: bed and breakfast inns, roadside stands, on-farm visits, local restaurants and bistros, farmers' markets, online sales, ethnic clubs, day spas. These are just examples of what can be done with herbs. The list goes on, and from those lists, there are also niche ways to assort herbs with other crops if one is so inclined, which turns the very same herb into a "new" crop: Cosmetic packages can include clary sage and lavender along with cured loofas and fresh cucumbers (known for skin health) and goat-milk soap. Herbs can blend with flowers to create "tussy mussies" with romantic ancient meanings. Italian herbs go along with Italian vegetables and monthly cooking classes on the farm. Salad dressing herbs sell well with salad greens.

Micro eco-farmers sometimes start with themes, such as "wild birds" or specific seasons. They may grow birdhouse gourds and birdseed, or autumn decorations like rare ornamental corn and gourds. They may create the community's favorite U-pick, self-serve flower patch, starting in spring with pink pussy willows and ending in late fall with hardy chrysanthemums. They may create body care products such as rose-infused skin toner and chamomile hair rinse, or assemble goods such as gift baskets. A popular gift basket on my own farm was "Gourmet Salad for Two," which contained at least two different heads of gourmet lettuce, at least one darker green addition such as a rare, very sweet spinach or Italian greens, heirloom "Lemon" cucumbers, "French Market" baby carrots, a rainbow of cherry tomatoes that ripened to yellow, orange, and red, and herbs like lemon basil, parsley and others for homemade dressings. It was topped with an assortment of edible flowers, a recipe, two spatterware salad bowls, two matching cloth napkins, and a bottle of herb-flavored vinegar.

"Walking Onions" at Five Springs Farm, a small CSA farm on 18 acres, with about one-half acre of intensively managed raised bed gardens near Bear Lake, Michigan. The CSA provides vegetables to about 25 families, a gourmet restaurant and a deli. Co-owners Jim Sluyter and Jo Meller, in addition to farming, produce a farm newsletter called "The Community Farm." "We are enthusiastic about networking with other farmers and farm members through this newsletter," they say. Contact them at csafarm@jackpine.com.

If the farmer loves flowers and he or she knows florists are already full of hothouse carnations and long-stem roses, the micro eco-farmer may instead cultivate big, sunny bouquets of butter yellow "Sole d'Oro" sunflowers, to be mixed with deep burgundy "Velvet Queen" sunflowers. In a year or two, they may add "Florence" cornflowers selected from the ancient blue field flower to offer pink, lavender, violet, white and blue. They may sell them in bouquets and U-pick plants, or even as edible flowers, as I did on Island Meadow Farm. But what if an aspiring farmer loves carnations and roses? They will offer what is not found in florists, perhaps "Fenbow's Nutmeg Clove Carnation," or "Sarah Rose" or "Fairy Rose" and "Mary Rose," and an ongoing supply of scented carnation and rose floral cologne, and perhaps summer classes on making body care products with flowers.

The microfarm may grow crops that could be mass-produced some day, but probably won't be. The Styrian oil pumpkin has been described as providing "olive oil from the vine." This crop could have a market not large enough now for a corporation to serve, yet just right for a micro farmer who knows how to draw enthusiasm with his or her own enthusiasm. Micro eco-farms are so adaptable that if their premium crop were taken over and cheaply produced, they'd just as quickly be on to something else. But this rarely happens to micro eco-farmers, because they attach add-ons to their crops which cannot be duplicated en masse. Their oil pumpkins might be marketed along with organic gourmet garlic and herbs to flavor the oils, and the distribution of home oil presses, with classes on the farm on making flavored oils. Instead, they may create massage oils from their own pumpkins and market them to their local massage school and day spas. Potential mass production is not that large a concern for the micro eco-farm, because their crops are just part of the "whole package" of who they are in either their local community or the target interest group they serve.

Selling the "whole package" along with the crop

Tomatoes, once again, and mushrooms are good focus points to explain the whole package that goes along with the micro eco-farm's crops, and

Crops from The Chile Man farm (see page 26) include figs (rt.) and Red Savina chiles (below).

to show how powerful this concept is to the microfarm's success and security.

Mushrooms are successfully grown by microfarmers in city basements and are also collected wild from microfarm rural woods, while heirloom tomatoes are grown on Toronto's city rooftops and delivered just-picked to local restaurants, and also grown on rural, small acreages. In the country, the wild mushrooms are sold via internet and at a local farmers' market. Also in the country, heirloom tomatoes are added to the offerings of a small country U-pick farm that lets visitors enjoy the pony and ducks (while manure from the pony and ducks provides the material for growing the country-grown tomatoes).

Customers of both the city and country tomatoes and mushrooms purchase them partly for what is also attached to them: One offers a taste of rural life and a direct experience with the flora and fauna realms and the enjoyment of crops known to come from wild conditions. The other offers instant delivery of just-picked crops for fast-paced city restaurants. Both the city and country microfarms offer basically the same crops, but they attach something else to each crop, filling different niches. Selling the farm's own special touch allows the customer to buy an entire experience rather than a commodity off a shelf.

When my customers subscribed to a weekly bag of produce to be picked up from the farm, I filled old-time bushel baskets with the produce and set them out in an old horse-drawn market wagon. Customers got a taste of nostalgia by picking up their produce in this manner. Jan Johnson, who owns a few acres in Washington State called Larkspur Farm, earns a full-time living for herself and children as a flower farmer. She begins her selling season in April when she invites other small farms and artisans to sell their wares on her property in an antique and garden show, then ends the season in November with a similar show. Customers can choose flowers direct from her gardens or purchase bouquets from her honor-system roadside flower stand. She has even allowed weddings to be held on her farm. Another small-acreage farm plants bedding flowers out in beautiful gardens with pathways lined with the flowers for sale so that customers can see them already growing in a beautiful garden setting. Customers can stroll along the flower-lined paths, choose a plant for their own garden, dig it up for purchase, and the farmer then will replace that empty space with a new blossoming plant.

A final experience for customers, which micro eco-farms share with all local ecological farmers, is their knowing that the farmer is selling a higher cause: The farm is healing the earth, restoring the strength of local communities, and preserving the diversity of healing, cosmetic and decorative plants. Regardless of the farmers' unique way of presenting the crops themselves, they may want to remember that customers of this century "want the farmer again." Often, connection to the farmer is also what customers are buying, and the crop or product created from it may be only a portion of why they enjoy circulating their money through the farmer.

Focusing and testing the market

The micro eco-farmers' challenge is not that there's nothing to choose from; it's instead focusing their choices from so many possibilities, then tapping into a natural market for them. If you're at your starting point, or revamping, you may want to begin with a theme or target, such as "only locally-adapted varieties" or "only heirlooms." Test market this for a season (or by word-of-mouth), then branch out from there. As a micro eco-farmer, you might collect possible market outlets for an ethnic theme, offering hard-to-find, sustainably-grown, ethnic crops, perhaps choosing Oriental, Southwest Asian, East Indian, Italian, French, Mexican, Native American, Early American or Regional American cuisine. From this focal point, begin a collection of crop possibilities. Offerings may include "Ambrosia" dill from Russia, "Violette di Firenze" Italian heirloom eggplant, and from the cuisine of the Piedmont region of Italy–the aromatic "Costoluto Genovese" tomato. Or, perhaps you'll choose Old English or American heritage crops.

A good way to ease into eventual full-time is to start by focusing on a specific season, or to integrate a microfarm segment into an already existing home business. If you choose autumn you may start out

Lion's Head Farm, near East Meredith, New York, raises porcelain hard-neck garlic, a relatively rare variety. According to owner Beverly Maynard, the hard-neck garlic has the best flavor, is long storing, and contains the highest level of allicin—one of the healthful ingredients in garlic. The photo shows the garlic being cured, which takes a few weeks. Contact: prinsol@catskill.net.

offering speckled "Swan" gourds, "Strawberry" ornamental popcorn (a bright-purple-red ornamental corn that can be popped), tear-drop shaped Japanese "Orange Hokkaido" squash for delicious winter soups and pies, bright yellow "Halloween in Paris" pumpkins and red-orange "Rouge Vif d'Etempes" pumpkins, both shaped like Cinderella's coach. To this you may add large white pumpkins and miniature white pumpkins for ghostly jack-o-lanterns or painted faces.

Perhaps you'll emphasize spring, instead, as a starting point. During 12 weeks in spring, an elderly couple earned $12,000 during that time from sales of their vegetable garden starts which they grew in their old plastic-hoop greenhouse. They sold them to people stopping by for their ongoing, year-round garage sale.

You may simply start as a hobby and watch it grow. On Green Hope Farm in Meriden, New Hampshire, Molly Sheehan and her husband Jim, plant gardens in the shape of circular rows, star geometries, flower tunnels and patchworks of unique specimens. They harvest their flowers to create flower essences, which are healing solutions made by infusing flowers into liquid. Their gardens and essences started as a hobby and then grew into a huge mail-order business.

In conclusion, the micro eco-farmer seems tapped into infinity. Change is his/her friend, whereas change in the corporate system is devastating. The micro eco-farmer siphons in crops and their growing methods from the same apparent source that is changing the values of the society that supports the farm. They can grow most anything in

Crops from Salt Creek Farm (see page 29): Emerald Oakleaf lettuce (left) and Purple Skinned Igenheimer potatoes (below rt.). Emerald Oakleaf lettuce photo taken by Martin Hutten.

their gardens and have much less invested in equipment created for single, specific crops. If a farm begins with heirloom vegetables, fruits and culinary herbs, little by little, healing herbs may be added. Then more healing herbs and less food crops may prevail, according to the farmer's growing interests and the customers he or she attracts. It all depends on which market the farmer taps into and enjoys being around. The microfarmer who produces blue, pink and purple fleshed, freshly-dug organic potatoes with skins so tender they must be handled by hand, may eventually cross over into locally-produced potatoes that originated in Usury Hara, known as "The Village of Long Life." This rural village near Tokyo produces "satsumaimo," a type of sweet potato; "satoimo," a sticky white potato; and "imoji," a potato root. The villagers there are known for long life and smooth skin into their 90s.

The Elliots, creators of Idle Thyme Farm, own 15 acres, total. "We have committed to setting five acres aside as natural prairie wildlife habitat, so that leaves us ten acres for our animals and our crops. We are having lots of fun with a great pumpkin patch and so many other diverse ideas," said Shelly. The Elliots survived an unprecedented blizzard during one of their first years as a young farming family, but they came back strong. They have remembered to have fun. Like all micro eco-farmers, they are grounded to the earth, with an ear open to humankind's next step or next phase of growth. Along with their home gardening and hobby-farm cousins, the inflow of possible crops and markets for them is abundant.

CHAPTER THREE

The Animals: Ancient Breeds, Wildlife, and a New Vision

The sheep are so content to live here and they add a lot to the farm just with their presence. The llamas are very friendly and are a real joy to have as pets. The pure joy that the people get from coming here to see the animals makes it well worth the effort to raise them. The Angora rabbits are for the fiber for the hats I make. The animals—especially the chickens—eat the compost from the kitchen. The llamas love to eat the vegetables that we grow as well.

—Mariam Massaro, Singing Brook Farm

Reuniting the fauna with the flora

Nowhere in nature do plants exist without their complement of animals ("animals" also referring to such creatures and earthworms and lacewings). One's waste is the other's sustenance and the cycle continues. Segregate one from the other, and an imbalance occurs that leads toward a downward spiral. Micro eco-farmers are aware of this union. There are many who choose to farm without animals at all; yet they allow their farm to beneficially interact with the local wildlife. Some install bluebird houses for insect balance or plant borders that attract predatory insects which in turn help re-establish the insect balance in their location. (See more on "Wildlife" below).

The older, multi-faceted farms of early America that incorporated farm animals with plant crops took advantage of the symbiosis shared between the plant and animal realms. On their farms, sharp beaks and manure were an asset that fertilized the crops and removed parasites from fields. These assets are instead huge expenses for egg factories that must remove the manure, cut off the beaks, and purchase artificial feed that the chickens once foraged on their own.

Nowhere in nature do plants exist without their complement of animals... One's waste is the other's sustenance and the cycle continues. Segregate one from the other, and an imbalance occurs that leads toward a downward spiral.

Micro eco-farms integrate animals again to restore the lost gains to their overall farm plan that came from that symbiosis. The resulting sellable product of the animals or animal product seems almost to be an extra bonus in some cases. But where products are concerned, more and more specialty and niche animal products that don't include the well-known meat, egg and commercial dairy commodities are being called for, such as grass-fed dairy products that are fermented right on the farm, or dairy and rare wool from animals raised in a natural setting. The vegetarian movement is stronger now, as well, which has boosted demands for non-meat products. (For a free pamphlet, *The Vegetarian Alternative,* on considering the health, humanitarian, environmental, and spiritual reasons for considering becoming a vegetarian—as well as many vegetarian resources—see "New World Publishing" in the Resources and Networking section.)

It is the intent of this chapter to explore the new and the niche beyond the already well-known meat, egg and commercial dairy products and to explore how farm animals are returning with new options, a new vision, and allowing micro-sized eco-farms to prosper in their presence.

Ecological minifarms have discovered marketing outlets and living conditions on small ground in which the animals thrive healthily throughout their lives in a manner that also enhances the larger ecosystem. It is my intent to show these minifarm living conditions and marketing outlets aside from the products you will ultimately sell from your animals. These microfarm models may be adapted by any entrepreneur for their own personal enterprise and belief system.

Filling the need for locally produced or specialty products

Micro eco-farms often produce products for markets that only they could know about locally and through shared-interest groups, which is why there are few national resources telling them of these markets. They weave one-of-a-kind luxurious throws from their Angora goats and rabbits, selling them to local bed and breakfasts and country inns, which in turn sell them to their customers. They provide fresh milk, honey and herbal ingredients for local holistic salons, offer a U-milk program through the community supported agriculture model, which has allowed customers to buy "shares" in the dairy animal, and therefore may allow them to use their milk raw or in any form they prefer. They raise a flock of rare-breed sheep and give spinning and knitting lessons, or package rug hooking kits for children. They make old-fashioned calligraphy pens from their poultry's molted feathers, and combine them with calligraphy workshops and garden composting workshops. They sell snowy-white wool from their horned Dorset sheep, white Angora goats or white Shetland sheep, and along with that package their natural plant dye crops. From dairies that pasture-graze their animals on mineral-restored soils, they create on-farm, processed, fermented dairy products in the form humans most benefit from: enzyme-rich, probiotic, lacto-fermented yogurt, cottage cheese, buttermilk, kefir, sour cream, butter, ghee or rennetless cheese. Other farm specialties include products like homemade ice cream flavored with the farm's organic strawberries, small-batch yogurt flavored with local wild blackberries, and ethnic, boutique or artisan cheeses. They collect rabbit and goat manure for a worm casting operation and sell the castings in bags at local farmers' markets.

Types of animals, and more product possibilities

Micro eco-farmers sometimes raise the rarer farm breeds that provide long-forgotten (or never known) products and services, which also reestablishes these breeds. The rare breeds often carry valuable traits such as the ability to give healthy births without human intervention. They produce well in natural conditions, are wonderful mothers, and forage their own food without the need for artificial feeding conditions.

Examples include hardy, rare wool and dairy breeds, such as tiny Welsh Mountain and Shetland sheep with their luxurious specialty wool, and rare miniature dairy goats and cows. On her three-acre Iron Horse Farm, Deborah Smith of Sherborn, Massachusetts has raised Shetland sheep since 1998. The Shetland breed dates back 1,000 years, and still retains its strong survival instincts of thriving on marginal land. The sheep are known for their soft, fine wool that was used to make the legendary "wedding-ring shawls," where selected neck wool was knitted into lace shawls so fine they could be drawn through a wedding ring. "And the colors!" Deborah said. They choose not to raise sheep for slaughter, but concentrate on the wool. Shetlands are the producers of 11 natural colors of wool. "I have a very good following for my fleeces and roving for hand-spinners," she said. "We have all colors of Shetlands here on our farm."

Dominique and Silver-laced Wyandotte chickens and Welch Harlequin ducks naturally free range. Alpacas, domesticated in South America more than 6,000 years ago, provide wool in 22 different colors. Milking sheep provide gourmet dairy products along with hand-spinners' wool. In the late 20th century, America alone imported 23 million pounds of sheep cheese a year. Satin Angora rabbits, a cross between the French Angora and shorthaired Satin rabbit, provide multiple colors of specialty wool that looks like spun glass and feels like silk. Miniature horses, donkeys and cows give humans experiences with animals and provide offspring to other breeders and small acreage owners, and raw material for farm and garden compost for sale.

On the really small end, endangered butterflies, with the blessing of the close eye of environmental organizations, are bred and sold to be released during warm-weather weddings and other celebrations. Earthworm ranches supply local gardeners with worms and castings.

This is just the beginning. Micro eco-farms also rescue more common farm animals and raise them in eco-conditions, turning their products into something that can't be found in the mass market. Because these animals produce from hormone-free, wild and remineralized pasture, and are allowed to follow their natural instincts such as twilight grazing and natural mothering, eco-farmed animals provide higher quantity and better quality products, such as full-spectrum mineralized CLA/Omega-rich sheep, goat, and cow's dairy products and finer, stronger wool.

One-of-a-kind products

Micro eco-farmers profit directly from their animals in three main ways: First, they sell raw material, such as feathers, milk, and wool, direct to artisans, local chefs or others who further turn them into yet other very unique products. This allows the farmer to get 100 percent of the retail price with no middle people taking a cut. Because the artisans are further enhancing the product for their own profit, they are able and willing to seek out and pay for the raw material that is not otherwise mass-produced, which allows their own crafted product to stand out from the rest.

For example, rather than raising large numbers of sheep to contribute to mass-produced standard-white, standard-length wool at 80 cents per pound, micro eco-farmers raise a few hand-tended specialty wool animals that produce natural colors and a variety of textures to sell direct to hand spinners and fiber artists, and rare-breed offspring to other small farms. On a group farm on a Washington State island, the farm owners sell "gentled" rare breed animals to other breeders as 4-H projects and as pets. The hand-tended fiber from their llamas, rare sheep and alpacas is sold raw, carded and ready to spin, and crafted into wearables. Upon a recent visit there, I held an eight-ounce bag of soft, carded alpaca wool which sold for 18 dollars.

Second, micro eco-farmers may have their own on-farm cottage industry or on-farm processing program which crafts products from their farm crops, as with the farm above which also sells wearables from the wool. A farming couple in Maine established a hand-made blanket company using only Shetland sheep wool.

In Oregon, Karen Cross and Ardena Lonien each had four dairy goats when they decided to start making goat-milk soap, which turned into a small business supplying ten states.

On Goat Lady Dairy of North Carolina, a herd of 35 – 50 dairy goats provides fresh milk for about 250 pounds a week of daily hand-made farmstead cheese. The owners, Ginne Tate, her brother Steve and his wife, Lee, keep three dairy goat breeds: the Saanen, the Nubian and the French Alpine. "We have about 15 acres in pasture for grazing our dairy goats and a couple of Jersey steers," Steve explained. "We probably have an acre for organic herbs, vegetables and fruits including composting areas. Thirty-plus acres is kept as wild areas in woods and stream. The rest is our three house sites, barns, a dairy building, etc."

Author's son, Jereme, harvests apples from young orchard with help from farm geese, ducks and Lucky, the dog.

The cheeses are crafted into several styles and flavors, including Chevre Logs, which are hand-molded and air-dried for a week, and "Smokey Mountain Round," which is a Goat Lady Dairy original. This cheese is formed and dried into a round, then smoked over pecan wood logs. More than 50 purebred goat kids are sold for family herds, and the Tate family follows the natural dairy cycle, which means milking for ten months, breeding in fall, and kidding in early spring. They call mid-December to mid-February their "Sabbath" time for all to rest, rejuvenate and prepare for the next cycle. They rent out a 1,500 square-foot meeting room, hold occasional farm open houses where a fee is charged per person to tour the farm and taste the cheese. All visits are arranged and they do not allow unannounced visitors. Monthly dinners at the dairy are offered in their new dairy barn.

Gail and Ed Cole of one-acre Ponders End Farm in Avra Valley, Arizona, raise and sell purebred Nigerian dwarf dairy goats, similar in size to the more familiar Pygmy goat, but a distinct breed from that popular pet goat. Along with milk and cheese, Gail makes luxurious soaps from her Nigerians' milk. "Nigerian milk is sweet and higher in butterfat than that of most large dairy goat and cattle breeds, which makes it ideal for making cheese and soap," she said. Although local bed and breakfast inns, guesthouses, and salons have requested this soap from Gail, she so far prefers to make it for friends and family.

Third, human-animal interaction has proven to be a very valuable "product" or service to a society that has removed itself from the natural world it once dwelled with. Studies continue to mount on the health benefits of human and animal interaction. This can be achieved on the farm without really trying. Customers enjoy the plant and other crops the farmer produces, while the animals are around grazing and putting on a show. Bed and breakfast inns have a greater customer draw when farm animals are about. On less than an acre, a U-pick (or U-gather) filbert farm allows peacocks to forage under the groves during certain times of the season. I watched as delighted children gazed at the peacocks and took home shimmery feathers found in the woodland-like setting (and asked to go back).

On nine-acre Bird Haven Farm, a single horse named Dusty mows the brush surrounding a U-pick blueberry patch along with a few sheep companions. They attract customers, and their manure contributes to the fertility of the blueberries. There's more about Dusty and Bird Haven Farm in the Chapter 5.

The human-animal interaction, though, can also be quite intentional and even the main source of income where school tours, petting barnyards, animal health workshops and ecotourism, or "edutainment" is offered for a fee. Revenue comes for farms and ranches that find humans healing from their interaction with tamed horses, llamas, bunnies, and other creatures. "Our ranch is really profitable," said a manager of a children's horse farm, "Parents tell me they're glad to support our farm instead of spending money on drug and alcohol rehabilitation for their kids."

A fourth "product" demonstrated in the above examples is raising offspring in natural conditions, making them "human friendly," then selling them to other breeders, rural families, other microfarms, programs needing gentle animals for human interaction, or the people needing just one or two for their craft products, such as the fiber artist with a couple of wether (neutered male) sheep. Goats, sheep and geese are also bred or rented out as forest manager assistants for brush patrol, working with the humans to remove brush that has grown unnaturally out of hand, rather than using herbicides.

"We have a waiting list for the breeding stock we produce, and I credit this to the fact that we strive to keep many different bloodlines in our Soay sheep, as well as an informational and helpful website for marketing them," said Ronda Jemtegaard of Greener Pastures Farm in Washington State. On five acres, she has chosen small, easy-to-care-for livestock, including rare, primitive British Soay sheep, American Soay sheep, American Buff Geese, Chocolate Runner Ducks, and a flock of heritage chickens. The primitive breeds she raises, which have never been bred for modern feedlots and artificial feed, are lighter on the land and need less, and often zero, medical attention. They produce on more natural forage conditions, turning weeds and brush invasion into wool and soil fertility.

Ronda's farm is an exceptional model of animal health and productivity on small acreage. She combines pasturing techniques described more in Chapter 5 and adds a source of the new spectrum of trace minerals explained in Chapter 4, (in her case, she uses kelp).

Here, Ronda gives a picture of her farm: "Keeping more than one species on a small farm is easy enough, and rotating the flocks (or herds) through small paddocks keeps the pasture healthy and lush. I liken keeping several species of small livestock on a small farm to Mel Bartholomew's "Square Foot Gardening Method." The pastures are used intensively, but the difference is that each pasture is given a rest in order to return lush and green again. This requires lots of permanent fencing, which is an investment in time and money only in the beginning. An alternative is the use of portable interior fencing, which requires the investment of time and patience during its daily use. I'm a 'lazy farmer.' I'd rather do it right the first time and not have to mess around with it again later. I know people with farms no larger than mine, and with no more animals than mine, who spend hours each day feeding and watering their livestock. By setting up the fences, barns, hay feeders, water tanks and such correctly the first time, it means that my daily feeding chores take no more than 30 minutes a day, and no time at all when the pasture is lush enough to provide their entire diet."

Animal farmhands, just for letting them do what they already want to do

Portable bottomless pens are used with pigs, poultry, rabbits and other animals, allowing lightweight moveable structures to generate "farm labor," while feeding the animals a no-cost, healthy diet. Marc Winterburn owns 12 acres, and concentrates on sustainably grown, mixed vegetables, with a specialty in chiles. He rotates his planting beds and uses chickens and pigs for cultivation and fertilization. Chickens are especially good for weeds and insects. "The pigs," Marc said, "are for their cultivation and fertilization expertise."

In open grazing, chickens may forage under orchards, getting a free, natural diet while eliminating fruit-eating insects, fertilizing the soil, eating weeds in the orchard, and preventing fruit diseases by consuming fallen fruit. Hogs weed and root under chestnut trees before and after chestnut harvest. Multi-species are grazed together for mutual benefit. Goats and cattle may graze the same pasture, both eating different plants—the goats eating the brush that would invade the grass and herbs, allowing for more herbs and grass for the cattle.

These micro eco-farmers who integrate farm animals into their system have discovered how, on very small ground, to raise disease-free, happy, profitable animals in natural conditions without taking up more land, while doing less work, and in a manner that mutually benefits the animals, the larger human community, and the entire ecosystem. This new vision of raising animals in a way that enriches rather than depletes the larger world is further explored in the chapter on Abundance Methods.

Wildlife

Hanna and Stella, Salt Creek Farm (see page 29).

Surrounding my Island Meadow Farm are hay fields, small family acreages and an organic cattle ranch. The coyotes were known to invade the fields and kill pets and farm animals. They were shot, and interestingly, when one went down, another filled its spot. That's now nature works. Below our hill, I stopped and talked to a neighbor while taking a walk. He also used to kill the coyotes to protect his flock of geese and ducks. But one day, he saw a female coyote with a broken leg, and had the notion to feed her instead of kill her. She recovered and remained wild, but as her natural instincts allowed, she now saw the man who had fed her as part of her pack. His territory became her territory. She saw the poultry he left at peace, and she did the same.

Running through his ducks and geese to eat rats and mice without harming the poultry, she also protected "hers and his territory" from other coyotes. So he relaxed back, allowing a coyote to protect him from coyotes. That's also how nature works if we choose.

While it can be devastating to feed certain wild animals in a manner that makes them dependent on hand-outs, there may still be more for humans to learn about how we can co-exist in a mutually beneficial way when the animals remain wild and respectful, and when a balance is struck between the shared environment and needs of both.

We may see a ceasing of the human-created problem of insect "pests," when we stop growing weak crops (see next chapter) that attract the insects meant to devour weak plants, and when we stop growing large mono-crops that do not allow the pests' own predators (birds, toads, predatory insects) to live nearby. Though unhealthy plants naturally attract insect predators whose job is to eliminate them to make room for healthier plans, all is still not lost for slightly weakened plants if there is an ounce of health and hope left in the plants. Near the end of the last century, studies coming from Illinois State University, Washington State University, and USDA research centers in Illinois and Florida revealed how plants under attack also call out to the *predators of the insects* that are attacking them. Plants can even tell the difference between an artificial human-made injury and an actual caterpillar bite. Signals are sent out to call in the prey of the unwanted insect, and if one plant sends out the signal, other nearby plants of other species will also increase their natural defenses as well. With huge fields of one mono-crop that do not supply living grounds for pest predators, and their natural plant companions nowhere near, their natural predatory controls are gone.

If you don't already live near a woodland or a neighborhood of natural diversity, beneficial insect borders or tiny, corner wildlife sanctuaries can draw in nature's guests for you. The Department of Fish and Wildlife in some states has inexpensive packets that help homeowners create small backyard sanctuaries. "Bug" baths and birdbaths will attract them as well. Bug baths consist merely of a shallow dish with stones and water, allowing the insects to get to water without drowning. Ladybugs are well-known aphid eaters, and lacewings do the same. They and other beneficial insects get a secondary diet from certain nectar plants and need shelter and water. Beneficial insect borders are being planted around gardens and small farms to attract ladybugs, lacewings, and the many pollinating insects. Garden books and websites list the specific mixtures for these borders and mixes are sold where garden supplies are found. Inverted clay pots with openings attract insect-eating toads, while wren and bluebird homes attract birds that eat many insects, and some birds will even drive away other fruit-eating birds.

Author's daughter, Elise, rides Dixie, the dapple gray pony of Island Meadow Farm.

Homes for pollinating orchard mason bees and bumblebees add to the other pollinators besides those from the insect borders or the farm's honeybees. Mulberry trees provide food for otherwise cherry-loving birds; while clover and mineral licks are loved by deer who have been known to choose them over human crops. While some animals may overpopulate human areas, such as rats and mice, many other animals—including those who reduce the ones we want less of—are naturally territorial. If their instincts are intact, they maintain boundaries with others of their species. Landowners have been successful at planting a supply of food for them without attracting an over-population.

Selling the whole experience

As we know now, customers are supporting more than just the product when paying an eco-farmer with animals. They get renewed connections and receive experiences with the animal realm they can rarely get anymore, enjoying once-a-year open-houses, ongoing farm visits or stopping off at the roadside stand. They know the animals are well taken care of and are thriving in a manner that helps the larger ecosystem restore itself.

Unique, premium products with a "farm personality," attract loyal customers, and in some cases, all or most advertising costs disappear, and the micro eco-farmers find themselves with customer waiting lists.

> *A friend of ours has five acres on which he raises goats, chickens, rabbits, geese, turkeys, and pigs—all on pasture. He's an excellent chef, but he quit working at restaurants because he was dissatisfied with the quality of the food available to him. He decided he'd have to raise it himself. His farm has been in operation for only two years, but already he supplies food for his family and a growing number of neighbors. People stop by George's farm store on the way home from work and buy goat cheese, pate, quiche, ravioli, soup, apple cider, fresh vegetables, and pasta sauces. People shop at George's first, and then go to the supermarket for the things his farm does not provide.*
>
> – Jo Robinson.

There is no other place like George's, and aspiring microfarmers can use his example to adapt their own unique version that combines their passions, beliefs and hobbies with farming, whether they are fiber artists, feather artisans, or ethnic connoisseurs.

As with plant crops, when the farmer creates a personal signature associated with a unique product, the farm products are rarely "stolen for mass production." For example, hand-harvested inner silken wool of naturally-colored Angora rabbits from sunshine-raised bunnies is difficult to mass produce. Commercial Angora is often machine sheared, harvesting the outer coat along with the softer under coat. Hand harvesting gathers only the silken inner coat that naturally sheds during the year. But even if assembly-line rabbitries bus in $2-an-hour migrants to hand-pluck caged rabbits, the citizens of this planet are becoming too sophisticated to not see or pay for the difference.

Beyond humane

The successful eco-farmer raises his or her animals in a manner that the specific animal species calls for according to natural law. While natural law would not have its creatures eating an artificial diet in dark confinement, it also does not call for a sheep to be fed from a pretty plastic bowl scoured daily by loving owners. The sheep may be better off lowering its head to the green earth, picking up beneficial bacterial during its dining. This is a new form of "humane" that taps into the earth's wisdom. Animals raised according to their own natural laws produce better with less labor for the farmer, ending the idea that humane treatment is a form of fluff that costs the bottom line. Elsie Horton of Elsie's Goat Farm in Texas has researched and written excellent material on raising miniature goats (see Resources and Networking).

"When we first started raising the little goats it was next to impossible to find reliable information about them," she said. "Unfortunately, the poor animals often pay the price when well-meaning but uninformed people decide to 'get a couple of goats.' I shudder when I think about how some of our first goats suffered from our 'loving' care.' "

And while it's good to start with basic information on each species before purchasing them, Elsie points out that what works for one farm may not work for the next. Realize that most farm animals don't read the books we've written about them nor necessarily agree with what veterinarians have to say, and understanding animals is an ongoing process. Plus, one flock or herd of animals will act differently from location to location. Elsie also eventually successfully raised geese after finding that city water was detrimental to their health. But she describes her first attempt at raising them here which also illustrates that, once animals are purchased, owners need to be prepared to place them in new homes should they change their minds.

"I will not relate to you here all of the hair raising experiences we suffered from this roving band of marauders," she said of the first geese she owned. "But, after they trapped a neighbor in her house one day, we finally had to get rid of them. She had to call us on the phone to come and get them. (We could hear the geese honking loudly in the background.) I ventured over there reluctantly with my usual weapon of choice—the kitchen broom—and a handful of corn.

"It appeared that the bunch of miscreants got bored with their one-half acre pond and decided they liked the neighbor's kiddy pool better. Needless to say, the neighbor didn't want the geese to deposit green silt in her pool and bravely tried to 'shoo' them away. That's when the geese went into attack formation. Luckily this woman was young enough to sprint to her front door, barely managing to slam it in time to keep them from following her into the house. And that's where I found them: still honking loudly, spreading slimy droppings on her front porch and viciously eating—or should I say ripping at—her yellow roses that had been adorning (prior to their visit) a wooden trellis alongside her patio.

"The neighbor's son had a friend who was the caretaker for a local lakeside park and wildlife reserve. The two young men came to our house later that day with cages and loaded 'our guys' into their truck. They promised us that the alligators would not eat the geese, and I guess they were right. About a year later I overheard an elderly gentleman at the store in town telling the clerk that some 'big ol' white geese' had chased him away from his fishing spot. Geese can live 100 years, so I guess my great-grand grandchildren will someday get a chance to visit them. As for me, hopefully, I will never see them again." Elsie's next group of geese and all her geese thereafter turned out very tame and well-behaved.

There are reasons why animals must chew, swim, graze, scratch, flap and peck according to their specific natural law. It often isn't until we take these actions away that we later find how they benefited both the animal's health and the farm.

When animals are eating humans, or humans are eating animals, there's an obvious reason why a distance is kept between them. African children are taught how to avoid being the lion's prey, and early American farm children, for emotional reasons, were taught never to name the animals raised for slaughter. In that sense, it was wise and understandable that creating pets out of farm animals was looked down upon.

But mutually beneficial companionship (versus being distant enemies) appears to also be encoded within universal law once the animals and humans have an understanding they won't be sharing each other for dinner. It does not appear to be merely human-created fluff. Native Americans are known to have once approached wolves in a brotherly fashion. Even when the "I'm not eating you" agreement is only temporary, the lion lies peacefully with gentle grazers when no meat-eating is about to ensue. For those whose agreement is permanent (using their animals alive on their farm for the lifespan of the animal), this human-animal relationship can be allowed to grow and be further explored, rather than purposely avoided.

In conclusion, each unique microfarmer who chooses farm animals weaves his or her own tapestry of what part the animals play in their own farm system. They find direct sales for most of the animals' products, and benefit from labor the animals offer.

"My animals contribute every day to the large compost pile near our round pen," said Mariam Massaro. "The sheep, llamas and horses all live together in a big pasture and barn recently built near my large gardens. I have the horses to ride and to brush and to make the compost. The sheep and the llamas are for fiber for my ponchos, blankets and sweaters (another small farm business on the side.) The sheep are amazing at keeping the brush down along the brook that was overgrown with wild rose bushes." There are many choices for adding farm animals to a small farming system beyond mainstream reasons, and both the larger ecosystems and outer community benefit as the micro eco-farm returns to the neighborhood again.

CHAPTER FOUR

Three New Foundations for "Restoring Eden"

This resulted very quickly and dramatically in freedom from all plant diseases, as well as hardy and highly productive crops. The cabbage worms disappeared in short order, and best of all, the cabbages grew like I had never seen before. I had the biggest, healthiest and most wonderfully robust cabbages I had ever grown. Many crops, particularly green beans, lasted and produced for a whole month longer into early winter.

—Peter Weis, 1.5 acre Spring Farm, Salt Spring Island, Canada

Humans sometimes think of Eden as a place where life was effortlessly and perpetually sustained by something greater than human fear and hard labor. We have continually been motivated to study methods that come closer and closer to producing this, and to study pristine locations on the earth that appear to come close to demonstrating Eden.

This motivation has spawned three farm restoration methods that have paralleled the organic movement and are now coming more into popular view. When utilized, the earth then accelerates completely on her own all human efforts of tilling, weeding, fertilizing, storing extra nutrients for later release, immobilizing heavy toxic metals, disease elimination, pest elimination, frost tolerance, heat tolerance, flood tolerance and drought tolerance. She'll provide, free of charge, all rooting hormones, blooming hormones, and growth hormones released to the plants at precisely the right times in precisely the right amounts. She will provide flavors and aromas and nutrition beyond anything the world has seen in millennia, and will provide more of it, with less human work, on smaller amounts of land than we ever thought possible.

Charles Wilber created a system of growing tomatoes that average more than 340 pounds per plant. Conventional 20th century agriculture reported an average of eight pounds per plant.

"Just a few of my plants will out-produce an entire field of conventionally grown tomatoes—with a fraction of the work," he states in his book, *How to Grow World Record Tomatoes.* "I hardly think of weeds as stress simply because I do not have any. My distaste for hoeing has put weeds out of the equation." Charles' ingenious system, described more in the next chapter, includes a combination of methods, but might be described as giving the earth what she wants from the top, and letting her do the work below the soil, undisturbed.

The three "foundations of Eden" are often simple, fast, and sometimes even free for the farmer to re-create. First, we restore all the earth's original elements needed for full creation, including a mineral treasury now known to be much larger than before. Then we restore the earth's original regenerative microbes (vs. degenerative or absent), of which there are so many that the human mind can barely fathom them. And third, we leave the earth alone. These three foundational methods can enhance almost all micro eco-farming, no matter what bioregion they produce in or what other organic or sustainable methods and ingredients they use on their farm. They eventually result in no-till, low-labor, disease and insect-free permanent growing areas that withstand frost and drought, and produce (in micro-spaces) a bounty of whatever you want to produce. They are adapted even to the farms in plastic swimming pools on city rooftops, basement mushroom growers, and somewhat to sustainable hydroponics growing. Once these foundations are in place, farmers then choose from the particular remaining abundance methods (see next chapter) that apply to their own communities and personal farm.

Restore all the earth's original elements needed for full creation

At the start of the 21st century, humans recognized that 92 basic elements are the building blocks of everything in creation. These 92 basic elements are all that the universe uses to create all form: rocks, plants, people. And the funny thing is, when the elements aren't there, the universe can't use them for its creations. It seems to want to produce life, abundance and health. But take away its elements, and disease and scarcity appear in life's place as it has to makeshift with missing elements.

Of these 92, 83 are used to make living things. The life force (sun's energy, cosmic energy) uses these 83 elements to create living systems, and then that life force generates itself through those living systems here on earth.

Peter Weis, quoted above, is now retired from his farm and educates and consults on health and sustainable methods that enhance current organic practices. Peter's family grew vegetables, herbs and flowers and operated a small 8 x 16 ft. greenhouse. Besides their own year-round supply of vegetables and herbs, they sold radishes, lettuces, spinach, carrots, beets, cucumbers, green peppers, tomatoes, zucchinis, cabbages, parsley, oregano, sage, savory, basil and dill at the farmers' market on Salt Spring Island in Canada. "We also sold vegetable, herb and flower bedding plants grown in our greenhouse," he said. He then went on to explain to me how we get from 92 to 83. "Some of these (original 92 elements) occur in stars but not on the Earth. Technetium, Promethium, Francium, Astatine, for instance, do not naturally occur on the Earth; they are very short-lived byproducts of nuclear fission. And Rubidium ignites upon contact with air and reacts violently with water. Then there are the "Noble Gasses"—Argon, Krypton, Neon and Xenon—which do not react with any of the other elements, and therefore are not needed nor used by biological processes such as plant and human life. So, it comes down to about 83 elements remaining for biological processes—and that's ALL elements, including such fundamentals as oxygen, carbon and hydrogen."

Yet we have been leaving out about 64 of these 83 elements. Here's how that happened:

The 83 elements are often categorized into a variety of lists describing them as "major," "minor," and "trace" elements. All of them are important, however. Some just take a lot of themselves to do their job, while some operate in minute amounts. All must be in balance with each other. Take one element out of context and the others cannot do their job. Take one trace element from the entire balanced spectrum and it, alone, can become a toxin.

One popular category-breakdown of these 83 elements is as follows:

—Three are considered "major elements." They are usually taken from water and air (hydrogen, carbon and oxygen).

—Now we're at 80. Of these, three more "major" elements are also needed in large supply (nitrogen, phosphorus and potassium), the familiar N-P-K trio noted on fertilizers and usually readily available in organically conditioned soil and live soils that can take some elements out of the atmosphere.

—Now we're at 77. Of these 77, five vitally important elements must do their job in balanced, slightly lesser amounts. These are the "minor nutrients" (magnesium, calcium, sulphur, sodium and chloride).

This leaves 72 vitally important elements that do their job in very trace amounts. Of these 72, eight (boron, copper, manganese, molybdenum, zinc, iodine, cobalt and selenium) are often maintained to some extent by conventional and organic farms somewhere along the food chain, either directly in the soil or in farm animal feed. This leaves 64 completely out of the picture. Because they are used in such minute quantities and used to be plentiful in our soils, it was easy to assume that no matter what we put back into the soil, they will be present. We are now coming to realize that has not been the case.

As the Remineralize the Earth Foundation explains, when it comes to the need for trace elements, the least can exert the greatest effect. The tiniest trace element amount can be even more crucial to proper growth and health than a major element in larger amount. Thus, the least is often the most.

Living things could also be described as electrical energy systems for life force to generate itself through. When all elements are in place in proper order, those electrical currents can run as intended.

But just as important is that when all of the elements are available in *correct proportion,* they can, in unison, appear to create miracles. If the elements needed to create a particular life form calls for large amounts of calcium, for example, *when all trace elements are available,* they can restructure themselves into the element that is most needed. This is a group effort. They must all be available in balanced form for this natural system to occur.

With only the major and minor elements available in the soil, and just a few of the trace elements, even they cannot do their job in perfection when the remaining trace element co-factors are gone. A four year study cited in *World Health and Ecology* in the later 20th century, and then reported in the title *Empty Harvest* by Bernard Jensen and Mark Anderson, showed calcium levels in crops dropped 41 percent, magnesium, 22 percent, potassium 28 percent and phosphorus eight percent over the four years, in spite of the fact that those elements had been restored to the soil. It revealed to these authors that the plants weren't even using the replaced minerals, such as calcium and potassium, adequately. The rest of the players (trace minerals) that activate these elements had not been invited.

Calcium will sweeten your apples, but not without its many co-factors, the most obvious being magnesium, which is sometimes not even considered a minor element. It is in very high concentration in the sea, but some consider much of the earth's surface now very, very low in this mineral. Dr. Norm Shealy, M.D., a world-famous neurosurgeon who gave up that practice for holistic medicine, states that every known illness is associated with a magnesium deficiency. And magnesium, like all the others, works in conjunction and must have its counterparts beyond its commonly known calcium relationship. When plants then get their supply of magnesium and pass it on to animal and humans in a form we can digest and assimilate, health can be restored. Further, we may find that trace elements play an even more important role.

High Prairie Truck Farm & Gardens, located near Franktown, Colorado, utilizes raised beds in straw bale containers. Hoop structures protect high value crops such as eggplant, tomatoes, peppers, etc. The CSA growing area is less than one-acre of actual planting space, serving between ten and 20 families each summer. According to Jude Wallace and Cathi Allen, co-owners of this small operation, "We have devised a system of straw-bale gardening with shade cloth and drip systems to enhance our short growing season at 6300 ft. altitude. The shade cloth protects our crops from the strong sun, hail, wind, some pests, and gives us a structure from which we can protect with plastic and row cover to extend our season." Contact: truckfarm1@aol.com.

When they were once plentiful

The earth was formed with all of the elements available and in balance. They were, and still are, everywhere, in various forms including large rock formations. Over time, wind and other natural conditions broke them down into tiny particles. When the particles became small enough, microorganisms could then get their sustenance from them, and then pass that full-spectrum of minerals on to more complex forms of life: the plants and animals. Then, the familiar cycle of life as we know it sets in motion. The debris of the plants and animals, once decomposed, returns to the soil the original elements in their tiny particle size, and new life springs from that. Humans, obviously, take most of these elements in balance into their bodies through plants and animals that get them from the soil. While there are some plants that can absorb all the elements needed from the atmosphere, humans are still maintaining their physiologies by consuming the fruits of the earth.

Where they went and how we get them back

Ironically, natural erosion can be seen as both a friend and a deterrent to life. Some believe it was natural erosion in the first place that broke the minerals down into small enough particles for life as we know it to begin. Once life got hold of that, and created a cycle of continually returning those small particles to the earth, life grew and flourished on earth.

But both human and natural erosion can also wash away the tiny particles, and if nothing ever returns them, the land remains barren, or at best, life in those locations is not optimal. Life exists, but with disease. Plant, animal and human life in those locations may be only half of what it could be if the full-spectrum of elements was available on a regular basis.

The most obvious forms of human erosion which removed this full-spectrum of elements are land development and the removing of plant and animal crops from a parcel of land for agriculture, and then not returning the full spectrum back. Further, humans extract and concentrate some elements out of their natural balance, such as aluminum. Then, during the 20th century, only a few (N-P-K and those eight or so recognized trace elements) were returned to the soil in chemical form.

Organic practices have done an excellent job of restoring the original cycle of life by composting and mulching, which returns the matter back to the soil. But today, we can't always be assured that all 83 elements are restored in correct proportion. Even when pesticides and herbicides are not used, manure and straw for compost did not necessarily come from fields or animals with access to all 83 elements. Traditional organic methods are very good, of course, at unlocking what remaining elements are available, unlike chemical farming. The trace element iron was shown in a study to be 80 times more plentiful in organic crops than conventional counterparts. But as far as continually restoring all of these elements, year after year, about the only way humans had accomplished this was when whole sea ingredients were added to the farming system. Kelp and organic foliar sprays from sea emulsions are gaining momentum in sustainable agriculture. Their balanced elements are a large part of their effectiveness. In the latter part of the 20th century, the University of Maine reported on a surprising experimental find that spraying sea-derived emulsion on potato plants protected them from aphids and Colorado potato beetles. A marine biologist involved in the experiment concluded that the plants were so healthy, the bugs just passed them by.

Today, the full-spectrum of minerals lay waiting to restore the earth. They lay in rock formations that are too large to be used by the microorganisms and they have been washed to the sea. As well, nature, in her brilliance, has plants ("weeds") that are particularly good at thriving in eroded areas, sending out roots deep into the earth to mine the missing particles, bring them up into their body structures, and then ultimately restore them to the topsoil once their leaves and other body parts return to the soil. Peter Weis had a particularly eroded spot on his farm that attracted the Canadian thistle, a weed in which much time and money has been spent on study of its removal. Peter cut the thistle; lay it down on its growing area as mulch and sprinkled it with chicken manure and seaweed. The thistle eventually restored the eroded area and Peter's additions accelerated the process. Once its job was complete, Canadian thistle never again returned to his property, now that Peter's soil was completely restored.

Currently, the two most popular and available methods for restoring this lost balance are rock dusts and sea fertilizers. Rock dusts are powders in which humans mine or pulverize mineral-balanced large rock into small enough particles to be available to the microorganisms.

In Michigan, 100 native maple tree saplings were sprinkled with a glacial rock dust containing a large amount of trace minerals. The remaining 400 were left alone. The next spring, the trees with the rock dust grew three feet in one spring spurt, while the others grew one foot. But what's more, they also grew noticeably healthier. They had darker color; the diameter of their trunks was larger; and their "leaf tatter" was minimal.

A separate company researched and tested their version of a mineral replenisher derived from an ancient clay bed mixed with marine minerals. Their studies showed improvements in plant quality, growth, maturity, and resistance to insects.

Organic sea fertilizers include kelp, and even seawater that has had imbalances removed. Peter Weis primarily used seaweed to restore what he describes as the 72+ trace elements (the 64 we've left out plus the 8 trace elements humans have tried to restore). It was free for the gathering near his Spring Farm, which had started out with very poor soil. "Our gardens were not only indefinitely sustainable, but grew richer and more fertile year by year," he said. "We got the complete natural spectrum of the 72+ trace minerals from a seaweed mulch throughout most of our gardens, and seaweed added to our compost."

Many gardeners and farmers have had faith that there is "just something very good" about using soil amendments from the sea. Peter set out to find out why and came to understand how vital all the elements play in the creation of life. He was on to something the world wasn't ready for yet. So was Julius Hensel, a nutritional biochemist who wrote the book *Bread from Stones* in the 1880s, describing rock dusts and their benefits. Others covered the subject as well, but it remained hidden from mainstream knowledge. Dr. Maynard Murray, a medical scientist, published a book in 1976 entitled, *Sea Energy Agriculture.* He cited experiments, projects and demonstrations that pointed to the slow disappearance of crop and farm animal diseases once the earth's original mineral treasury was restored. Yet his work still wasn't ready to be accepted at the time. We now know even more than the doctor then understood, and this type of work is now picking up momentum. Experiments, new companies and new organic growing methods are beginning to reestablish the elements in a variety of ways, while finding answers to concerns about ocean pollution at the same time.

Disease and hardship seem only to be a second choice for the earth once an imbalance occurs. And that leads to how we created "pests" and "disease" in the first place. When an animal is dying, the vultures and coyotes close in even before the actual death. The dying animal is sending them a signal through the airwaves. As nature's "cleaners," the vultures and coyotes will then quickly dispose of the body. In the plant world, when a plant is weak from lack of all of its nutrients, disease can set in. In weakness and disease, it sends out signals of distress. It, too, calls to nature's "cleaners." Pests will arrive to clean up the weak plants, leaving the healthy ones to thrive. The pests will proliferate according to how much of a food supply they see on hand.

Along with monocropping (the growing of acres and acres of a single crop), the loss of the earth's natural mineral balance appears to be one of the causes of insect pests. They don't come out in large numbers nor do damage until we provide them with a feast of sick plants. "Fast growth" that comes from providing just one element, such as nitrogen, will call the aphids in by the droves. The backstage trace elements that would have allowed the fast growth *along with health and vigor* are missing. Accelerated growth from the full-spectrum, rather than nitrogen alone, also keeps the plants healthy and more free from aphids.

Weeds appear generally to restore eroded soil and protect bare ground. Once the soil has been returned to life and balance with a healthy crop to hold the topsoil, weeds don't have to show up for work. Both pests and weeds are something farmers and gardeners can see less and less of as their soils return to original life and balance.

In summary, a first step in creating a lush garden or pastures for animals is choosing a method to re-establish and maintain the full spectrum of soil elements from a source that is already in perfect balance and contains them in small enough particles. Sources include a yearly application of rock dust or adding kelp to the soil or compost. Some cover crop mixes with far-reaching roots may mine enough trace minerals also, depending on the geographical area. Sea-derived foliar feeds add a quick surface boost while the soil is being restored. In a United States field test for a company supplying organic fertilizers, grapes sprayed with liquid kelp produced 35 percent more per acre than grapes not sprayed. Farm animals can be supplemented with kelp or other forms of full-spectrum minerals while their pastures are upgraded. Their manure will then be a better source of restoring lost minerals to their own pastures (or the farmer's compost). When your soil is re-established, your crops and animals then own what Dr. Maynard Murray terms "a complete physiological chemistry."

Restore the soil's original regenerative microbes, the earth's digestive system

The earth is, by nature, teaming with microbes. All of them have their beneficial place in nature, but there are certain ones that we call beneficial and those we call pathogens. Those who study the science of microbes have concluded that beneficial microbes were once in far higher numbers when the earth was pristine. Only when their numbers fell did the pathogens come to be seen as a problem. When kept in check by the beneficial microbes, even the pathogens have a valuable place in keeping the earth and its inhabitants healthy. Their job appears to be to remove and destroy organisms that are not healthy in order to make way for healthier organisms to take their place. But when they go unchecked without a higher supply of beneficial microbes, and when they are allowed to mutate even further, they are believed to assist in deterioration of even healthy living systems.

By restoring the original balance of beneficial microbes both above and below the ground, life has been shown to return in a most optimal way. Art Biggert of Ocean Sky Farm applies beneficial microbes with several compost tea formulas. He reports that this prevents downy and powdery mildew of squash plants, late blight of potatoes and tomatoes, botrytis of grapes and basil, leaf curl of peach and plum trees, anthracnose of beans and black spot of roses. Weeds often grow in less than optimal soil and disappear once the soil is alive with microbes. Art also used to have a buttercup infestation on his tight-clay soil. By applying compost tea once a month, the buttercups decreased significantly without any herbicides.

Blue Moon Farm is a one-acre farm on Vashon Island in Puget Sound, Washington, that specializes in fall and winter vegetables. In the spring and summer Blue Moon Farm grows a gourmet salad mix and a handful of spring and summer vegetables for sale at the local farmers' market. It also operates a winter CSA from mid-October until the end of the year. When the CSA ends, customers can continue to order a limited variety of the hardier crops throughout the winter. Most crops are grown on permanent raised beds using both biodynamic and organic practices. Contact: bluemoonfarm2@earthlink.net.

Photograph © Ray Pfortner. Ray Pfortner is a professional photographer specializing in environmental issues from pollution and recycling to habitat restoration and land use. Contact: pfortner@centurytel.net, or www.RayPfortner.com.

Beneficial microbes are needed to allow all the minerals described above to flow from the soil to the plants in the precise amounts and timing. They are the plants' digestive systems. The plants can then convert sunlight to plant energy. Art Biggert explained, "Plants do not derive their nutrients from the soil. They eat from the second table after the microorganisms process each other and the carbohydrates in the soil. The microorganisms' metabolites (waste products) are nitrogen, rooting hormones, blooming hormones, growth hormones, etc. They compete with and feed on pathogens. They also provide a food source for larger critters such as worms, beneficial nematodes and the like."

Life-giving microbes both feed our crops and keep the disease-causing microbes away. Of the microbes that cover the earth, only two to ten percent are reported to be potentially disease causing—and even these serve the universe as a sort of second-choice method for decomposing a life form that is too imbalanced to continue.

When we give the disease-causing microbes plenty of imbalances to grow on and remove the conditions needed for the beneficial microbes to keep them in check, we create an oversupply of pathogens. Then we try to "kill them" (along with any remaining beneficials). Then they mutate. Then we get into trouble. Restoring the conditions needed for the beneficials to return is what can get us out of trouble.

As reported by the Remineralize the Earth foundation, conventional agriculture tried to bypass this full-spectrum mineral/beneficial microbe system by using synthetic chemicals to supply a chosen few elements as soluble salts that are directly absorbed by plant roots. But they explain that many beneficial bacteria and fungi actually pump nutrients into roots *ten times or more faster* when needed by the plant than when chemicals are absorbed by the plants' roots. Besides "fast," the release is also steady, ongoing and gradual when a slower speed is more important to the plant. The precise amounts of nutrients needed at the precise times of growth for each individual plant are released when the microbes are living among and on the plants' roots to do the job the universe created them for.

Further, a mineral-rich soil full of beneficial microorganisms maintains a sweet soil with a perfectly balanced pH, and is more likely to maintain soils with life force that the plants take into their systems. The microorganisms build on themselves and the soil gets richer and richer each year. Conversely, chemicals create an over-acid condition, which kills what beneficial microorganisms are left in the soil and gets less fertile and more and more "dead" each year. So, the plants then appear to need even more chemicals the next year, and more the next. They get weaker and weaker, calling to their predators and succumbing to disease, and therefore appear to need even more pesticides and chemical disease killers. Further, the humus that the microorganisms created from mulch and other dead matter once acted like sponges that held water and nutrients during rain and released it gently during drought. As this humus disappears, the plants appear to need even more water; so more water is pumped onto them. The soil can no longer sponge up excess water, so what isn't immediately sucked up by plants runs off, leaching more of the nutrients, causing a need for even more chemicals and more water.

By now, the behind-the-scenes benefits the trace minerals and microbes provide, such as added frost and drought protection, have long since gone. The plants, with all that chemical nitrogen, may look "big and green" but they are calling the bugs in. With zero natural predators nearby and no natural mix of other plants for the bugs to have to hunt through, the bugs are getting ready for the feast and orgy of their lives, with lots of babies on the way. Now, it appears we need more pesticides. Meanwhile, all over the earth, the two to ten percentage of microbes that cause disease are proliferating in the absence of their cousins which were supposed to keep them in check.

When the earth's original mineral/microbe system is restored, your farm can flourish as it was originally intended on its own. Grace Gershuny and Joe Smillie, in the fourth edition of their book, *The Soul of the Soil,* expand on the benefits of humus, which occurs as a result of a living soil: Humus holds mineral nutrients in an extremely efficient manner and keeps them from being washed away in rain or irrigation. It immobilizes many toxic heavy metals. It maintains a perfect pH balance in the soil, buffering either over-acidity or over-alkalinity. It can hold 80 – 90 percent of its weight in water to release it when the soil becomes dry.

"The plain truth is, beneficial microorganisms form the basis of our food chain. There are five million micro-organisms in a teaspoon of healthy soil," said Art Biggert.

Most organic growing methods have always relied at least somewhat on beneficial microbes. That's what composting and mulching are all

At Riverbend Gardens, near Fayetteville, Arkansas, vegetables (spring lettuce and other greens, peas, and onions, cherry tomatoes, sweet and hot peppers, etc.) and specialty cut flowers (daffodils, tulips, bearded iris, Louisiana iris, peonics, ornamental onions, glads, tuberoses, zinnias, sunflowers, etc.) are sold at the local farmers' market. Contact: riverbend@ipa.net.

about. Sustainable farmers of this century are finding ways to restore them even quicker, and in even larger numbers in the form of accelerated compost, foliar sprays, and "live water" (see next chapter) that appear to be reintroducing microbes that once flourished on our pristine earth. Applied above ground, they are eliminating disease and keeping barns sweet smelling.

When we also give the microbes their entire spectrum of trace and major minerals as described above, and then offer some of the other remaining 83 elements such as oxygen (as in air when we turn our compost piles) and hydrogen (as when it rains) they thrive. Earthworms and humus return to add air chambers to the soil, which keeps the air supply available even in flooding rains, which further allows the microbe population to continue and expand its colonies. Trace minerals such as boron, as reported by Grace Gershuny and Joe Smillie, are then able to do their work of protein synthesis, starch and sugar transport, root development, fruit and seed formation, and water uptake and transformation. University studies show that only a thin top layer of compost on soil significantly reduces damage from major plant diseases.

In summary, after you've secured your full-spectrum mineral source, create or enhance an existing compost, and/or utilize a mulching or foliar spraying system to allow even more of the microbes to return and thrive in optimal conditions. There are many well-known organic composting systems, mulching systems, manure tea and foliar spraying programs that can be enhanced with the full-spectrum of minerals to help return beneficial microbes. There's a system for all personalities and all current certifying requirements of your location. Farmers can purchase micro-organisms, compost tea makers, or study the science and generate their own supply until conditions for microbes become self-sustaining.

Leave the earth alone

Once your spot of earth has what she wants, you can then stop doing her job for her with hard labor. She'll stop the disease, build and aerate the soil and eliminate the weeds for you. For the most part, she will do a lot of the irrigation for you as well.

In nature, the 83 elements are used and recycled without continually ripping into the soil. There is an established micro-universe down there, where delicate webs of life continually construct miniature roadways and passages and cooperative systems. As just one example, the thread-like beneficial soil fungi, "mycorrhizae," establish themselves onto roots to supply plants with nutrients. Their filaments then extend out from the roots' ordinary reach to bring even more nutrients to the plant. They open the soil up for oxygen and water. Year after year, these beneficial chambers and associations build on each other, getting more and more complex and valuable. Once dug up, this is all disturbed, to say the least.

Even giving soils too much air, as in rototilling, kicks in the activity of some microbes too quickly, allowing them to starve when the supply of litter isn't available for them to break down. Accelerated aeration is usually best left for tumbling compost systems that continually refeed the microbes in concentration in a small composting area. Art Biggert cautions that even spraying on too many beneficial microbes directly onto soil without enough litter for them to break down can be a waste of time.

Nature, instead, allows nitrogen and carbon in the form of leaf litter and animal manure to drop only to the surface without it being forced into the ground. The earthworms and other microbes come to the surface and bring this debris down into the soil, which helps create perfectly balanced air channels. This is how Charles Wilber grows tomato plants that reach more than 20 feet high. He replenishes mostly from the top, only. And as Ken Hargesheimer, who teaches Permanent Bed No-Till Agriculture and mini-farming all over the world, explains, "Permanent bed agriculture is the answer and is being used by many people new to agriculture as well as established, innovative farmers."

There is often an initial tilling in the beginning. But once established, when it comes time to turn in a cover crop or remove crop residues such as squash plants, consider the heavy mulch techniques that simply smother them. Ruth Stout pioneered several "leave the soil alone and work less" systems. They are published in several of her books including *How to Have a Green Thumb Without an Aching Back: A New Method of Mulch Gardening*. Pat Lanza invented the "Lasagna Gardening" system, where she layers mulching materials to create a "no-toil, power soil" system. She not only doesn't disturb the garden bed, she doesn't even disturb the sod or weeds where the garden bed is going to go before the garden bed even exists. She smothers it with newspaper, piles her mulch layers on top, and even plants in them immediately. Charles Wilber avoids tilling also to protect the delicate roots spreading from his tomato plants. Another alternative is chicken tractors, in which chickens are moved over planted beds in bottomless pens, shred remains while they drop fertilizer and remove insects. Those with larger growing areas may find machine digging necessary at least once per year, but the longer they can allow the universe beneath their crops to establish, the more they can benefit from that soil community.

In summary, choose your planting areas, whether they are pots on your porch, an artistic design of circular raised beds, or a "Permaculture" garden (see next chapter), so the soil beneath can remain as undisturbed as possible. Then, separate from these, create a source that generates organic top dressing that contains the full-spectrum of minerals.

The closer your piece of land comes to Eden, the more your crop and animals' potential can return as originally intended. Plants grow lush and faster. Fruits sweeten. Colors deepen. Customers can't always articulate just what they taste from these crops. It's subtle. It's been a long, long time since they've experienced this flavor. But somewhere in their genetic or collective conscious memory, they remember Eden.

Adaptations to this include seeding the beds after production with a deep-rooted mineral-mining cover crop that also supplies nitrogen, cutting and laying it as mulch, and perhaps topping that with seaweed. When animals are involved, another adaptation is rotating planting fields with animal pasture with a system that rototills only once per year with lightweight tractors. For permanent animal pastures, farm animals can be rotated amidst smaller paddocks to produce lush pastures without reseeding, as is explained further in the next chapter.

To sum up this chapter, "Eden's soil" is re-created by beneficial microbes interacting with all 83 elements *in balance* needed for life, which are recycled from the top without disturbing the soil's living underground universe.

One thing I admire so much about micro eco-farmers is their ability to be on the leading edge and succeed with what works now, today, but with always an eye for what's coming tomorrow without jumping in too quickly. A new field of study suggests that the microbes we call pathogens can, themselves, transmute into beneficial microbes. How this occurs and how we can tap into its potential is still being studied. But the closer your piece of land comes to Eden, the more your crop and animals' potential can return as originally intended, where blossoms and fruits come early and plentiful. Plants grow lush and faster. Crops thrive in heat and drought. Growth continues longer, even into frosts. "Pests" retreat and pass you by, and disease pathogens have no place to thrive. The original full spectrum of minerals is sent out into the world from sales of your crops, while the original condition of the earth's top layer in the area you are stewarding gets richer with every season. Precious water that lands on this soil is retained and naturally purified by the work of the beneficial microbes and their humus and minerals. Fruits sweeten. Colors deepen. Your crops (which are already varieties chosen for flavor, lovingly hand-tended and picked fresh from the vine) are full of "the flavor of life," where all minerals are restored and the life force and electrical system of the plant is in full swing. Customers can't always articulate just what they taste from these crops. It's subtle. It's been a long, long time since they've experienced this flavor. But somewhere in their genetic or collective conscious memory, they remember Eden.

"You should have seen our beets," said Peter Weis: "almost as big as soccer balls. We had vegetables coming out of our ears, as they say. This was followed soon thereafter by complete freedom from colds and flu in our family."

CHAPTER FIVE

The Abundance Methods: A Sampler of Productive Micro Eco-Farming Techniques

The best part of farming is experimenting and trading ideas. The trick is not to get so attached to one idea that you ignore other ways to make it even better or use energy trying to make your way "be right." I see that too much in farming.

—Market farmer, overheard at farmers' networking program

When farmers independently learn from a variety of systems, they uncover what you'll get a taste of below: Charles Wilber's previously mentioned 300+ pounds from a single tomato plant; a thriving nursery on 1/20th of an acre; deserts awakening again; 80 percent less labor with many times the yield per square foot; 100+ percent increase in income per square foot; 67 – 88 percent reduction in water needed for production; and herds and flocks of healthy, happy, profitable animals on very small acreages that maintain the local wildlife and restore the earth's biosphere.

With one foot deeply grounded in their land, and an ear to the next horizon, micro eco-farmers don't seem to get stuck with any one method and are continually integrating new innovations into their own personal system. If they consciously choose to stay with one system, it's usually from a perspective that there are other great choices as well, but this "choice," not "stuckness," happens to serve them well. They don't seem concerned about discerning the right way from the wrong way, but are uncovering natural law according to their own particular talents and interests. When farmers are independent—or perhaps we should say interdependent—and network with the whole world, their upward spiral moves at an incredible speed.

> *Micro eco-farmers don't seem to get stuck with any one method and are continually integrating new innovations into their own personal system... When farmers are independent—or perhaps we should say interdependent—and network with the whole world, their upward spiral moves at an incredible speed.*

This chapter gives a quick overview of methods that show huge potential, and the Resources and Networking section at the end of the book lists how to find out more about them. You'll find quite a mix. There are hints leading to the secrets behind ancient, lush paradises that surrounded spiritual castles and pyramids; methods proven in the American public eye over and over again; and systems that just didn't get picked up yet by the mainstream for one reason or another. Yet, quietly, off in their corners, they have been demonstrating that miracles are actually a very common phenomenon in nature.

The abundance methods are overviewed here for you to compare and see what interests you most before you dive into the vast amount of luscious possibilities out there. You'll find some are further described in books, associations, apprenticeships and workshops. Some have on-line groups and video productions. Choose one, or allow one to inspire you and create your own version. Or, choose five, and see if you can get "five," when synergized, to add up to "ten." Remember that as you read, they are not only all improving, but ten more no doubt are being invented somewhere in the world and this is only a sampler of those that have crossed the path of this author.

"I use no pesticides, herbicides, machines, fossil fuels—I deliver to homes and businesses by bike," said Aaron Brachfield, who farms as a full-time job from his apartment as well as on two rented lots from local homeowners. He is also a student at Colorado State University. He donates ten percent of his income to students of all ages to help them benefit their community in the form of technology, scientific, artistic, theological and philosophical grants, "in the hopes of reestablishing ethical capitalism and democracy." Aaron integrates a number of sustainable techniques into the unique whole that he is creating and is continually improving. One of the techniques he uses is Square Foot Gardening.

Abundant farming designs

Square Foot Gardening. This popular method of specified spacing available to the micro-sized farm started as a home gardening system and evolved into an additional system for backyard home businesses. It is now spreading across the world as it continues to improve along the way. Arranging the plants in a simple grid system grows large amounts of continually-harvested garden produce. Its inventor, Mel Bartholomew, found that conventional agriculture, as well as typical backyard gardening, wasted huge amounts of space and also created added labor to maintain that wasted space. Mel's system, he found, saves 80 percent of the space, time, and money involved in the conventional American home garden. All thinning, most weeding, and a lot of watering are eliminated as well. Structures built for it are very simple and he keeps the soil live and air-filled by laying down old, recycled boards as walking paths. "The best part," said Mel, "is that it will work for a part-time income of $10,000 to $15,000 dollars or a full-time income of up to $30,000 dollars just during the growing season—all in your own backyard and sold in your local neighborhood."

Intensive wide-row spacing. Closer spacing obviously means more production in the same amount of space. Many gardeners now know that wide rows for walking in and skinny rows for crops have it backwards. When you put six skinny rows together, you have a wide bed without room taken up between every one of those six rows for walking. Wide rows for planting and paths just wide enough to walk in are the method of wide-row planting. This even works for nurseries. Michael McGroarty (see Resources and Networking) grows more than 5,000 plants in his 1/20-acre nursery by arranging them in wide beds instead of thin rows. Your fertile soil alone allows more crops to be planted closer together without any specific spacing formulas. Roots don't have to spread as far in search of food and water. One gardener demonstrated

Mel Bartholomew, originator and author of Square Foot Gardening, sitting beside a 4 ft. by 16 ft. display garden. Each square foot (created by a visible grid) is planted with a different crop in a home garden—for a cash crop you would plant a single crop in each 4 ft. by 4 ft. block.

"Square Foot Gardening is a uniquely simplified method of gardening that produces 100 per cent of the harvest in only 20 per cent of the space," Bartholomew says, "and without all the hard work and drudgery of single row gardening. It can be done in as little as 4 ft. by 4 ft., or as large as you want." Website: www.squarefootgardening.com.

Lettuce planted in the "square foot gardening" method: This one shows four lettuce plants in each square foot.

how three stalks of corn grew from one planting hole in his rich, fertile soil. John Jeavons' method, and others below, elaborates on this with more specific systems.

Intercropping. This is similar to companion planting (see below) with the goal of harvesting more than one crop in the same planting bed to increase yields. As one example, a grower in Prairie Farm, Wisconsin, harvested a bumper crop of peas and beans by planting two eight-foot lengths of fence with both crops. Peas went in first when the ground thawed; and when they sprouted, the beans went in. After harvest, the peas were pulled as the beans were getting larger, and continued into the season. The famous French Intensive farmers grew radishes, carrots, lettuce and cauliflower simultaneously in one bed. Each matured at staggering times, allowing earlier crops to be harvested before larger ones grew enough to shade or crowd them out. Combined with vertical growing (see below) the harvests in a small area can be further multiplied. John Jeavons and Mel Bartholomew elaborate on this, and a growing number of books and public demonstrations describe intercropping in further detail.

Nook gardening. Planting in moveable wheelbarrows, hanging gardens, containers, and so on increases the production of micro eco-farmers. A small farm in Washington State supplemented their regular farm income by starting garden starts on their windowsills to sell at their market. Baskets of herbs can hang from fruit trees, chives and mint can grow around the edges of fruit trees, and greens can grow between raspberries.

Succession planting or continual harvest. This is the ongoing harvest that may start with radishes and greens in spring, fill in with carrots and tomatoes and cucumbers in summer, then winter squash in the fall. It's a system that micro-eco-farmers can benefit from since they don't invest in large machines to tend one specific crop and harvest it at only at one specific time. More often, micro eco-farmers grow crops for a continual supply of mixed crops, fresh-picked all season long. This is how they can sell the more common crops to eager customers. They don't have to find a thousand people who all want beans at the same time. Instead, they attract the customers who want a small portion of beans along with salad greens and French round zucchini and tomatoes and cucumbers, just a little of each, every week, and picked fresh. Their harvest is "slow-release" rather than the rush of tractors and seeding all at once, months of maintenance with no cash flow, and then a rush of urgent harvest.

Vertical growing. Mel Bartholomew of Square Foot Gardening, and Charles Wilber, author of *How to Grow World Record Tomatoes,* and others employ the idea of going up into the air instead of spreading across into the acreage. Both Charles' and Mel's systems are multi-faceted, but vertical growing by them and many others is proving very beneficial. Some of the most popular vertical fruits and vegetables are rambling or vining crops, including melons, cucumbers, sweet potatoes, tomatoes, pumpkins, and other summer and winter squashes. When they go up, they reap the benefits of air circulation and more sunshine. Every foot up is more harvest for you that does not take acreage space. Their fruits are often more blemish-free. Harvesting, whether that is by the farmer or the U-pick customers, is much easier with this method. Even when pruning the crops somewhat as they climb, Cornell University found that yields per square foot were consistently doubled when grown vertically.

Permaculture and the "layering effect." Permaculture strives to establish more and more bounty in harmony with nature, as the human does less and less labor. One component of Permaculture could be seen as another form of vertical growing forest style. Nature demonstrates layers of plant sizes that mutually benefit each other and some farmers are following this pattern. Tall trees (such as taller nut trees) are planted at certain intervals. Then in their dappled shade, filberts or other smaller trees that thrive in these conditions form the next layer. Around the edges where there's more sun, berries are planted as the next layer. At their edges, intensive wide-row gardens are grown as the bottom layer. In the shade of the filberts and taller trees are herbs, fern greenery, root plants, worm beds, mushrooms or other crops that thrive in these shadier conditions, as well as chickens or other fertilizer-producing and foraging small animals. Crops are interspersed and planned in a manner that does not interfere with harvest of the trees. This short description hardly begins to describe, however, the "between the lines" benefits of the synchronization that Permaculture offers. There are contacts in the Resources for further study on Permaculture.

Regionally adapted crops. Loss of crop diversity from 20th century practices is a hole that micro eco-farms are refilling. When only a handful of corn varieties are bred for certain bioregions and certain markets, then offered throughout the nation for all climates, all soil conditions, all harvesting methods, and to sell to all markets, yields of

Midwest-adapted corn, for example, will obviously be less when grown in the Pacific Northwest.

When, instead, a crop is adapted to a local bioregion, its production increases. The Resources and Networking section lists seed sources for both regionally-adapted crops and open-pollinated crops that allow the farmer to save his or her own seed year after year to allow them to acclimatize even further. Nature loves to adapt and diversify. If a hot-weather loving tomato manages to produce a ripened fruit in the cool summers of the Pacific Northwest, and its seeds are saved, the next generation will be slightly better. There are now countless varieties of tomatoes that grow well in the Pacific Northwest, where several decades ago, people relocating to that area were warned, "Whatever you do, don't try to grow tomatoes there."

On only two acres of natural woodland, William and Merry Fischer operate a tree farm, Rosario Ever Green, where they also process worm castings made from rabbit manure collected three times a year from professional rabbit breeders. More than 25 tree varieties thrive in the native forest nursery, which is allowed to grow mostly undisturbed. Trees and worm castings are sold at farmers' markets, yearly master gardeners' sales, and the annual Festival of Trees.

Flower garden of many varieties provides cut flowers in small space.

Companion planting. This is where crops that mutually benefit each other are planted together for increased yields and quality of both. It's hardly new. The natives planted their corn, beans and squash together for mutual benefit long ago. Today, however, more and more beneficial relationships are being discovered. There are combinations that enhance the flavor of herbs and help other plants deter their host insect predator. This science is described in many fun, classic books with new information coming out all the time.

Multi-cropping. One disadvantage of growing several crops mixed together in a single bed is the possible added harvest time needed to "search" among crops rather than sort of mindlessly picking one specific crop, which can be quicker. To overcome this and to take advantage of both companion planting and the classic crop rotation most farmers and gardeners employ, microfarms break a single planting bed into blocks. Each block grows one specific crop, but that crop grows next to its beneficial companion, and is moved to that new section during the next planting phase. It's a choice each individual grower has to make. There are several models, and John Jeavons (see below) describes one well in his literature.

Giving the planting areas a jump start

Permanent raised beds. A fertile soil, if left alone, will eventually become loose and crumbly and full of air chambers. Raised beds hurry this process along with the premise that it is okay, just this once, to disturb the soil. Often made by mixing native soil with compost, raised beds are loose planting beds a few inches higher than the surrounding ground. These often go along with wide-row planting and may be rectangles, circles, ovals, or whatever fits the location. They drain excess water faster and are warmer than the surrounding ground. Those in very hot climates will sometimes reverse this, creating sunken beds that stay cooler and retain moisture better. In *Lasagna Gardening,* mentioned in the previous chapter, the author Patricia Lanza describes her version of creating loose, lightly raised garden beds without ever tilling. She lays wet newspapers on the new growing area (even over sod); then layers (as in lasagna) mulching and compost materials; then plants immediately. A loose layer of new raised beds is created while growing the crops at the same time while the raised bed's soil improves as the season goes on. There are now many books and demonstrations on a variety of ways to create your first raised beds. Ken Hargesheimer (see Resources) teaches workshops based on permanent raised beds for gardening and mini-farming worldwide.

Biointensive. John Jeavons created a blend of the productive French Intensive method and the Biodynamic method (see below). In his book, *How to Grow More Vegetables: And Fruits, Nuts, Berries, Grains, and Other Crops Than You Ever Thought Possible on Less Land Than You Can Imagine,* he describes how he goes even further than creating raised beds above the ground. He also immediately creates air-channels in the subsoil below with a one-time hand-digging method, then adds a fluffy, fertile mix above this which becomes raised beds. This is just one part of his growing system, which has continually improved and is being used across the country and world. His system has been found to create a 100+ percent increase in income per square foot and a 67 – 88 percent reduction in water needed for production—and this is just the tip of the iceberg. His entire program is under the umbrella of his Ecology Action Foundation. (See Resources and Networking.)

Worm castings. The worms will come to your land. They can't help it. And as they eat mulch and manure and lawn clippings, they excrete "castings." Worm castings are known to unlock trace elements and micro-nutrients and sometimes to have more than ten times the amount of nitrogen, potassium, calcium, magnesium, phosphorus and potash than regular topsoil or compost, releasing these nutrients to the plants in a perfect timed-released fashion. The castings are loaded with live, beneficial microbes that are also believed to pull additional nutrients from the atmosphere.

Most organic farmers encourage earthworms. If you want to concentrate their benefits quickly, worm beds can be created to harvest worm castings in large amounts. There are many methods for harvesting worm castings. The Resources and Networking section will lead you to several.

Microclimates: increasing yield and extending the harvest

Nature creates pockets where extra heat and moisture are condensed. Ancient royal gardeners famed for high production mimicked this trick with good results, producing warmer weather crops in their cold northern climates. This allowed the kingdom to eat just-picked fruits, herbs and vegetables without shipping them in. Today, gardeners and farmers create microclimates with cold frames and greenhouses. These technologies are available everywhere, from make-your-own instructions, to kits, to buying completed units. Below are a couple of adaptations to the creation of microclimates that utilize the ecosystem approach to increasing abundance.

The American Intensive appliances. Leandre Poisson and Gretchen Vogel Poisson invented a system to extend the season and increase production in the same-sized area with simple, low-cost easily moveable "miniature greenhouses," which they call "appliances." Very lightweight

Blue Roof Organics grows vegetables and herbs on raised beds which it sells at a farmstand and farmers' markets near Stillwater, Minnesota. Soil fertility is achieved through the use of cover crops and nitrogen-fixing green manures. According to owner Sean Albiston, "I'm trying to develop living mulch crops on the beds and a mowable green manure between rows to be harvested with a vacuum setup for compost/mulch." Website: www.bluerooforganics.com.

but sturdy, the appliances are moved over the crops instead of moving the crops from the greenhouse. Further, they found they enjoyed what they call an "open bed" instead of traditional rectangle raised beds to get the most out of their growing space. Their book, *Solar Gardening: Growing Vegetables Year-Round the American Intensive Way* is listed in the Resources and Networking.

Four-Season Harvest method. The experienced organic farmer, Eliot Coleman, invented this system which may be adapted to the small farm. Fruits, vegetables and herbs are grown all season in temperate climates with inexpensive mobile greenhouses and other techniques. One difference in this system from the one described above is that Eliot utilizes mobile walk-in greenhouses along with the smaller sizes. *Four-Season Harvest,* by Eliot Coleman, Kathy Bary and Barbara Damrosch is listed under Resources.

Farm animals and pastures

As explained in Chapter 3, eco-farmed animals thrive in natural, outdoor conditions including outdoor pasture if this is their normal habitat. This produces a more valuable product—happy and healthy animals—and when done in tune with nature, benefits the larger ecosystem. Here are methods that can further improve natural conditions of animals and the profits made by the farmer.

Paddock rotation combined with species diversification. With these two systems, pastures that once held only a few animals per acre are now holding many, with pastures so lush and healthy they sometimes have to be mowed, whereas before the fields had become exhausted.

Paddock rotation springs from studying nature's most bountiful genesis. In nature, grazing and browsing animals "herd" and migrate from one area to another, leaving the first behind to completely recover in their absence. They also intermingle with other species. They eat their favorite plants, biting into them only once, and then move on. At the same time (or shortly after they leave) other animals are eating different plants as well as possible parasites from the grazers, thus making sure those plants don't take over the grasses and herbs of the grazers before they fully recover and the grazers return.

If the grazers and browsers were to remain in the same spot continually, biting into their favored plants over and over again, those favored plants would weaken and the less favored plants for grazers, with no other animals eating the less favored ones, would grow, spread, and take over. Parasites would build to disease-causing numbers. Yet this is the description of the typical 20th century pasturing system. When this pasturing method didn't prove economical, this system further deteriorated into feedlots, where animals stand around in stalls, are fed questionable grains grown in large megafields, often mixed with chemicals and dead, diseased body parts.

Eco-farms avoid both systems and look to new designs. Andre Voisin was a French farmer, chemist, biologist, and teacher at the Institute of Veterinary Medicine in Paris. He designed a plan to make nature's more optimal design work on farms, and farms of all sizes are benefiting from his system. Pastures are divided into smaller paddocks. The animals are rotated from one paddock to the next. When moved out, the former pasture recovers completely with no continued grazing, as is allowed in nature. When only a few acres is managed this way, it produces many more times the pasture feed for many more animals than the older system which exhausted the same amount of land.

Maggie and Richard Krieger of Saltspring Island Llamas & Alpacas in British Columbia have had herds of more than 40 llamas and 30 alpacas. They have transformed a four-acre open pasture to the Voison system to assist in their herd production. Their barn is in the center of the four acres, and radiating from it are 16 "pie slice"-shaped paddocks, each approximately one-quarter acre.

According to a report by Jill Heemstra, a Nebraska extension educator, successful rotational systems can vary from a simple two-pasture rotation to using 30 or more paddocks. She noted that some farmers are concerned about the cost of starting a rotational grazing system and suggested the use of inexpensive, portable electric fences as a way to test the rotation before more expensive permanent fencing is installed. As with any other cost, it must return a profit, which it does by eliminating weed-control costs, improved and more valuable products from the animals, and reduced dependence on harvested feed.

Managed grazing can work on any size piece of land and improves the environment, the health of the animals and people who work with them, and is also more profitable. On less than five acres, animals are often integrated with other crops, or on-farm products are created from the animals' wool or milk, for example, to generate additional income.

To further enhance the paddock rotational system, farmers intermingle animal species that mutually benefit each other with their grazing patterns. Any rancher will tell you that sheep eat the grass down too low and therefore make it unavailable to horses and cattle; but there are other species mixes that are very productive. This can be as simple as allowing goats and chickens to graze with or shortly after horses or cattle. Goats eat the tougher plants and "weeds" the cattle and horses prefer to leave alone, making room for more tender pasture for the horses and cattle. Chickens spread the large animals' manure and eat their parasites, flies and larvae, and wind-blown weed seeds. If chickens or goats are not part of the farm's actual profit system, a couple of pet miniature goats or retired laying hens can be rescued for free and given homes in the pasture. Do keep in mind the need for all forms of life to receive their full spectrum of minerals. Goats have particular mineral needs that the pasture may or may not provide, and they and chickens can benefit from trace mineral supplements like kelp, and then further spread those minerals through manure onto the land.

Herbal pastures. Animals, even natural grass grazers, eat a variety of plants, not just a simple mix of grass and legumes. Herbs planted with the pasture mix benefit the pasture and the animals that graze them in multiple ways. Medicinal herbs concentrate certain minerals and other nutrients in their plant bodies and are instinctively sought out during times of lactation, seasonal shedding, growing a winter coat, or stress of any kind, to create optimal health for the animals. In the books *Fertility Pastures* and *The Complete Herbal Handbook for Farm and Stable* (mentioned more below), both authors report horses, cows, and other farm animals thriving with no need for medication of any kind. Their books list possible mixtures that can be sprinkled into pasture mixes. The animals' instincts lead them to the precise herb necessary.

Natural prairies support many more plant species than the average commercial pasture mix of the last century. When pastures contain natural herbs like deep-rooted dandelion, balm, borage, fennel, fenugreek, wormwood and others, they draw up trace minerals that in turn feed the pasture, and they keep the farm animals healthy and productive. Newman Turner, author of *Fertility Pastures* (available from *Acres USA,* see Resources) points out the difference between cows grazing on a simpler mixture of grass and legumes, and cows grazing on an herbal pasture that also includes chicory, burnet, yarrow, and others. Milk yield, he reports, increases every time the cows return to the herbal pasture. This experiment went on for four years, year-round, and many variables were added for test purposes. But no matter the time of year or length between recoveries of pastures, the herbal pasture allows the cows to produce more in a healthy, natural manner.

Hedgerows, pasture trees, and the "European Long Acre." Once more, animals—including grass grazers—eat even more than wild "pasture." In the wild, they nibble here and there on leaves, bark, berries, fallen fruits and the like to round out their diet. Sometimes they even go to the seaside to nibble on seaweed. In Europe, healthy animals

Seedlings at Zestful Gardens (see page 32).

were tethered to one end of a mixed hedgerow and gradually moved to its other end. By eating herbs, tender grasses below the hedge, and hedge trimmings, they fulfilled all their nutritional requirements. As they were moved down, the first sections recovered fully before the animals returned. By the time they reached the end, they were brought right back to the fully-recovered beginning of the hedge again. The hedges divided property and grew along roadsides, yet provided an entire diet for the farmers' animals on small ground, or no ground at all. They came to be called the European long acre.

Plants that create productive hedgerows include those that offer good fence barriers, such as rose and honeysuckle, intertwined with hedge plants that offer leaf and twig forage, bark, nuts, seeds and fruits from winter through summer, and with the ability to grow unattended and to spring back quickly after browsing. Examples include hawthorn, willow, alder, chestnut, linden, poplar, maple, and some smaller fruit trees. The added benefit to the ecosystem is that the hedgerow provides a place for wildlife to live and eat as well.

Beneath the hedgerows, herbs as those described above under "herbal pastures," can be planted to reseed and return on their own, and to thrive in this moist, protected microclimate. Both *Fertility Pastures* by Newman Turner and *The Complete Herbal Handbook for Farm and*

Stable by Juliette de Bairacli Levy, describe hedgerow plants and the herbs that grow well beneath them.

When mixed hedgerows are then added to rotational paddocks as the fences that divide those paddocks (or in some other manner), even more food and health can be supplied to the animals on small ground. Then, adding trees and shrubs that don't compete with pasture grass but provide shade and extra food for the animals, further rounds out the animals' diets, allowing the farmer to bring in less from the outside during summer drought and do less work as far as storing and feeding hay. Locust trees, for example, are considered trees that do not compete with grass. They generate a highly-concentrated feed in the form of "bean pods," providing bushels of them each season. Persimmons and apples drop their fruit free for the eating, (some animal species should have fruits chopped first.) The Siberian pea shrub fixes nitrogen for surrounding pasture and fruit trees, while it also provides a favorite food for poultry. Willows grow profusely, and grazing and browsing animals enjoy their branches in summer drought when pasture growth is slower. Trees have also been discovered to conduct energy and collect moisture from the atmosphere, which benefits the meadows growing around them.

Portable, bottomless pens, and combining animals with orchards and vineyards. Chicken tractors, rabbit mowers and pig rototillers allow animals to move to new, natural territory to consume a free diet while serving the farmer at the same time by debugging, fertilization, etc., as described in the section on animals. Sir Walter, a gardening couple's pig, was moved along in a bottomless pen to do his particular form of rototilling. Chickens are moved in similar pens through pastures or over finished crop beds in a device called a chicken tractor. Yet other farmers use pheasants and rabbits to mow and debug grapevines. This provides labor for the farm while providing a free diet for the animals, which are far healthier and happier and more productive than commercial or grain-fed animals. Farm and garden networks and periodicals regularly report and swap new ideas on the idea of bottomless, portable pens and grazing animals in with other crops.

The deep litter method, or "scratch run" for poultry. When it's more productive for a microfarm to keep its poultry in one stationary location, here is a healthy method for doing this. Poultry, unlike rabbits, goats, sheep and other traditional farm animals, are more scavengers, and are able to gain health when eating some foods that are in the decomposing phase. In fact, they seem to benefit from the microbes as they break down raw material and release amino acids and other substances the poultry thrive on. However, to be healthy, the decomposing needs to happen with beneficial microorganisms, rather than those we more often associate with disease, rot and slime. Air and dry matter will help create the right environment for this.

The deep litter method is described briefly in *The Complete Herbal Handbook for Farm and Stable.* An outdoor run is built with a roof to keep out rain, while open-air fencing allows insects and fresh air to enter. Dry matter such as leaves, hay, straw, dried yard prunings are piled into the run, and then daily greens, garden scraps, lime, nuts, fruits, kelp and other offerings further described in the book are tossed into the run. This gives the chickens scratching exercise while they shred and mix up the dry material and keep it aerated. The litter creates an excellent compost material and only needs to be changed once or twice a year.

Accelerating the fertility of crops and pastures above the soil

Foliar Feeding and Compost Tea. Foliar feeding is the spraying of ecologically produced plant food onto the plants above ground. The use of compost tea is the watering of plants roots with a similar liquid solution.

Plants can absorb nutrition through their leaves' pores and roots in liquid form. The challenge with this so far as we know it, is if they are always fed intravenously in this manner, their relationship with soil microbes does not develop, and the rooting hormones and timed release nutrients, water holding ability of the soil, and future replenishment of the soil through microbes is not possible. The crops "blindly" absorb whatever is sent to them in the fluid whether that is more of what they need, less of what they need, or something they'll need tomorrow, but not today. Also, the soil fertility eventually deteriorates.

The chemical farming industry was counting on this form of feeding when they applied a few elements to their crops sometimes in liquid form. As we already understand, this proved to be the wrong direction to take.

Sustainable farming does allow, though, for foliar feeding and direct liquid feeding of the roots when in conjunction with building the soil, and when that soluble liquid is balanced with all elements and teeming with beneficial bacteria that devour and compete with the pathogens. The recipes for creating your own compost or manure tea are plenty in organic farming and gardening books of your choice. Keep in mind that some older recipes rely on the fact that manure sources were not from commercial, chemical-laden stables or dairies with an oversupply of pathogenic agents, and that new cautions may be in force. Eco-companies are also producing foliar liquids and compost teas that contain sea products and high concentrations of beneficial microorganisms. Look closely at the source of ingredients and the ethics of the company.

Water technology. Holy water, living water, thin water, alkalized water, and other names for restructured water are showing promise. They are not new, but rather the human mind now seems better able to discern them.

In some cases, technologies and discoveries appear to be re-creating the pristine, perfectly pH balanced water that once quenched our forests, farms, gardens, orchards and fields. In the 1950s, an inventor, James (Jim) Francis Martin, developed an odorless "live water" by fermenting seawater. This water, according to *Acres USA,* captured the entire spectrum of major, minor and trace elements in a mix with beneficial microbes. The water was witnessed to "cleanse polluted water and soils," and "cause the desert to deliver abundant crops." His water, actually, could also be put in the "foliar feeding and compost tea" category above, as microbes played an important role. Jim and a partner named Floyd Leland started a company to distribute this water, and named it "Medina Holy Water." Planes flew over dead and disaster areas, drop-

All life forms, including our own bodies, can be compared to electrical systems that draw life's energy into our bodies. As a radio with the precise elements in the right balance conducts radio waves, so, it appears, do living things conduct life energy when all elements are in place.

ping "Holy Water," where it allowed fields to flower and crops to grow. The conventional thinking of the time could not embrace how this could happen, and it was, it seems, an invention before its time. But slowly, more and more research by young professors is confirming the work of Martin. Several companies, including the Medina Agriculture Company, have since picked up the technology with aims to offer it again.

"Thin" and alkalized water is created in a number of ways, including putting small, specially-charged beads into the water supply which is claimed to change the water's pH and molecules to a more youthful and absorbing quality. My own observations are that this water looks and feels like morning dew. Other water supplements, when added in only small amounts to irrigation water, so far appear to accelerate life in gardens and farms.

Judy Hogge and her family own a 15-acre sustainable farm and are following some of the progress of these alkalizers and water enhancers, including one called the "Neutralizer." The studies are just beginning, but says Judy, "It appears that four ounces of soap (a special "soap" formula created by the same company as the "Neutralizer") and four ounces of Neutralizer in 25 gallons of water sprayed three days in a row on an area equals 2,000 lbs. of lime to sweeten the soil. Also, when you fertilize, so far it seems to be showing you can use 50 percent less than you would normally use." The Resources and Networking section lists a source for these water enhancers, and the number of sources is growing.

Inviting the entire universe

There seems to be a new way of looking at the planet. It appears the earth has friends in very high places. She's part of a cosmic union, not a fragile and separate thing that has no greater support from other influences. Seeing the planet as separate from this union, some feel, limits the service we offer her and ourselves. There are fascinating sciences in the field of atmosphere and planetary influences if they call you to study and apply them to your micro entrepreneurship.

Radionics/Paramagnetics. Yet another fascinating field is similar to the study of invisible radio waves and invisible electricity. Radionics and para (small) magnetics (the study of magnetic fields in nature) focus on the invisible energy that appears to conduct a form of waves or current to bring life to the planet. Why was the land around certain stone towers and temples so lush with paradise-like gardens? What type of "antennae" was used to create this? Why do certain spots on the planet feel so live and lush? Certain soils and environmental situations appear to "conduct" this energy more than others. Why?

When a soil is rich in all of its minerals and live with beneficial microbes, we can understand the chemistry side of how the plants and animals have everything they need for optimal life. But where does the "life" part of it come in? Certainly, many would say "God" and that's all there is to it. Yet those who study this field might query further: "What is God asking us to create more of, so that more of this life-energy, more of 'God,' if you will, can be conducted down to earth?" As many are becoming aware, all life forms, including our own bodies, can be compared to electrical systems that draw life's energy into our bodies. As a radio with the precise elements in the right balance conducts radio waves, so, it appears, do living things conduct life energy when all elements are in place.

As far as soils are concerned as conductors of this energy, it gets right back to those original 83 elements in balanced form again, coming together in unison to conduct the correct energy. And the studies continue to unlock even more reasons why certain soils conduct this energy on a higher level. The elements have north and south magnetic properties, negative and positive charges, and when in perfect unison—when the copper hooks up with the silver and the negative hooks up with the positive and everything returns to an original Eden-like balance—the circuitry for conducting this energy is opened wider. Soils and water, as with some of the waters described above, regain a pristine perfection. These balanced elements, as we've learned, are in the sea, as well as tied up in rocks and glaciers on the land. Rachael Carson is attributed to knowing how important the balanced elements are to living things, including humans. She told the world that we all carry in our veins a salty stream where the elements are combined in almost the same proportions as seawater.

Philip Callahan, PhD, author of many titles on sustainable farming and the study of soil, stones, and their conducting ability, has traveled the world with his wife to study the soils and stones and stone temples, and the mystery of the magnetic forces behind them. He states that fertile soils with all elements in balance can conduct this energy well. *Acres USA* (see Resources and Networking) regularly reports on the progress of this science, and offers books on the topic. This field may offer some fascinating answers in the future, including how pathogens may mutate to beneficial microbes, and how a mineral can transmute to another mineral when enough life force is available to the parcel of land.

Biodynamics. The biodynamic method orchestrates soil fertility, crop placement, the larger ecosystem, and the practice of concentrating life energy all into a symphony of bountiful, sustainable production. It was espoused by Rudolph Steiner, who lived and worked in the late 1800s and early 1900s. As of this writing it has its own certifying and training program. Workshops and literature are worldwide, and many gardens, microfarms and small farms have benefited from this program.

In conclusion, methods for returning the neighboring small farm and eco-entrepreneurship have been incubating and growing all over the planet for many years, and are surfacing now as technology and the human mind swap, understand and embrace them. They allow the neighboring farm and micro businesses to produce a higher quality and higher amount than previously thought by the mainstream. Food and other fruits of the earth can be produced in a manner that ignites your own interest and enthusiasm, which will invite even further exploration. They allow you to mix and blend and come to your own conclusions.

The micro eco-farmers' passion and innovation is, after all, the greatest abundance method the earth could ask for.

Robert Farr, "The Chile Man," owner of a ten-acre sustainable farm, and a member of the farmers' network mentioned at the beginning of this chapter, shares with readers of this book what he wrote for his June column in *American Farmland Trust Magazine.*

"Our sustainable approach involves more than just organics. We ensure healthier plants by incorporating Biodynamic techniques, and sowing and transplanting according to the cycles of the planets. We build topsoil by adding animal manures, compost, cover crops, and mulch to our raised beds. We companion plant, bringing in good bugs to devour the pests. We avoid synthetic fertilizers, herbicides and pesticides; even organic toxins throw the natural world out of balance. We interplant vegetables with herbs and small fruits (elderberries and figs provide a rich, loamy soil, and their dappled shade offers a mid-day respite from the sometimes brutal heat of Virginia summers)."

CHAPTER SIX

Orchestration, Diversity and Adaptability: The Power Behind the Micro Eco-Farm

This year we'll pick nearly 2,000,000 peppers (from 67 varieties) and produce more than 40,000 bottles of all-natural marinades, bbq sauces, mustards, and salsa in a "commercial" kitchen we built on the farm. Products are sold at festivals, online, during farm events, and in specialty food stores. We grow many other ingredients found in our products: gooseberries, elderberries, basil, cilantro, parsley, horseradish, green tomatoes. We also grow more than 100 varieties of vegetables, small fruits, and flowers to keep the farm in balance.

—Robert Farr, "The Chile Man," adapted from his column in *American Farmland Trust Magazine*

The idea of "selling one thing well" seems to apply to successful micro eco-farms in a new way. They sell one thing: their own unique "overall picture." From this central picture they branch out to an assortment, or diversity, of crops, products and services. The power behind this lies in the invisible ties between a diversity of living things and resources already in place. With the central picture as the unifying theme, they continually orchestrate all the diverse branches into a unifying whole that's worth more than the sum of the branches. With their ability to change with speed and grace, their adaptability allows the overall picture they're painting to improve continually.

In this chapter, I'll offer a sampler of micro eco-farms that prosper by using orchestration, diversity and adaptability. You'll see how they make a single crop worth many times its worth by how it benefits other crops or how it can be crafted into something besides the fresh-picked item off the farm. There are those who synchronize and unify what could appear to be many unrelated fragments, and those who actually specialize in a single crop yet are operating in the successful, diversified, micro eco-farm model. More market outlets available to the micro eco-farmer are revealed in this chapter as well.

As we have seen in past chapters, non-local (or centralized) corporate agribusiness has omitted markets that are commonly referred to as "niche" markets. A whole spectrum of resources, marketing opportunities and even a new vocabulary has emerged to fill the pockets corporate agribusiness cannot fill. "Vintage varieties, rare breeds, hand-tended, organically grown, locally-grown, on-farm processed, farmer-direct, ethnic, bio-regionally adapted:" the list is almost endless and continues to grow as micro farming evolves, creating many choices for non-corporate agribusiness. And if any one of these niches alone does not provide enough profit by itself, that can be even more valuable to the micro eco-farmer.

Since mass-production cannot profit from these niche products, yet citizens desire to purchase them, the micro eco-farmer can weave several of them together to add up to the required amount of income, while each crop or market somehow mutually benefits the other. An orchestrated farm that raises both rare-breed wool sheep and organic apricots reaps more on-farm resources than one that concentrates on just one of these. It acquires free mowing of the orchard and, therefore, free feed for the sheep. The interaction of the sheep and the orchard further provides free fertilizer along with a host of beneficial soil microorganisms and the disease control mechanisms that can come from the organisms.

Beyond the farm, bringing the rest of the symphony into the picture

Micro eco-farmers often get full retail price for their crops because they market directly to the citizens or artisans who will create a further product, such as with direct chef deliveries, farmers' markets, or U-picks. This direct farmer-customer contact allows an "invisible" benefit. During apricot harvest time, the sheep are grazing next to the orchard, and the U-pick customers experience animals up close. The farm then acquires a strong customer draw. This is not an experience customers can get at the supermarket, and it's one they may never forget. The whole farm, remember, is the greater reason citizens visit the farm, and the greater the experience, the more word-of-mouth spreads.

Orchestrating a unique diversity of crops, or crops with animals near one's local community can make what once seemed impossible, possible again. Can horses be valuable on small farms? Even just one? John and Silvija Pipiras of Bird Haven Farm in Southampton, Massachusetts, have balanced a family farming system to include a horse, a few sheep, blueberries, peaches, Asian pears, raspberries, and cut flowers along with on-farm processing of some of these plant crops. On their nine acres, the horse, Dusty, and sheep graze land surrounding the 1,300-plus high bush blueberries, creating a buffer zone between the cultivated fruit and the woods. This keeps intruding grass and brush away, and minimizes would-be hiding places for wildlife that may want to consume the crops. Some of this grazed land is inaccessible to cultivation or machinery, so what could have been wasted farmland has been put to production. The farm animals provide a strong customer attraction for the U-pick blueberry patch, creating a natural petting zoo surrounding the patch, an experience that cannot be offered by on-line grocery delivery, and it is certain to be relayed to family and friends long after the visit. Bird Haven Farm is selling: "I picked blueberries in the sunshine today just like I did when I was a little girl, and Shawna got to pet Dusty again."

On this farm, visitors are allowed to dump their leaves for a very modest fee, which saves the visitors' money. These leaves become bedding for the horse and sheep, dispensing of the hauling and purchase

Yarrow blooming at The Chile Man farm (see p. 26).

of animal bedding. Manure mixed with the leaves eventually fertilizes the fruit. The soil around the blueberries is now so fertile and full of humus it rarely needs irrigation. The fee from the leaves helps pay for the once-a-year tractor rental. So, if we isolate just Dusty, how much is he worth in advertising, brush control, mowing, wildlife patrol, soil fertility and irrigation?

John and Silvija further diversify with on-farm processed products (also called "value-added" products) which are products created from the farm's own crops. Silvija arranges stunning bouquets from their cut-flower crop. When fresh fruit ends around mid-September, the farm stand is filled with their own "Grandma's Best" jams and jellies made from surplus blueberries, peaches, and Asian pears.

Bob and Bonnie Gregson, authors of *Rebirth of the Small Family Farm,* created their full-time farm on about an acre by blending a mix of plant crops, farm animals, and on-farm processed products. They point out that calculating the financial return of adding farm animals to the picture as though they were isolated from the rest of the farm does not reveal their real value. As mentioned before, the chickens increase

Fertile Crescent Farm is a small family farm organically growing perennials and cut flowers on 1.5 acres in Northern Vermont. Products are sold at farmer's markets beginning in May with field-grown perennial plants. Soon afterwards, cut-flower production begins and flower bouquets are harvested and arranged from a mix of over 100 unusual, cottage garden flowers.

According to owner Susan O'Connell, "I focus on offering high quality flowers and a true connection with the farm. My customers have known me and my family for years, and I think one reason my bassinet flower display works so well is not only that it is cute and funny, but connects them once again with who we are. Many of my customers know the kids who outgrew that bassinet and know it is not only a display, but a statement about my child-rearing days!" Contact: fcfarm@pshift.com.

soil fertility just by following their own natural instincts, and as well, their city customers love seeing the chickens which further increases financial return. Blending several crops to draw customers in for more than one item allows diversity to do its magic. During a recent market farming network conversation, a farmer was concerned that honey just wasn't selling. Another farmer said they couldn't keep enough of it at their farm stand. The latter sold it along with other crops. The group conclusion was that people often won't make a special stop just for honey, but will stop to explore an assortment of choices. When they see honey along with other products, and then get a closer look at the locally-produced fruits and vegetables, and the gentle manner in which it was produced, they see honey, fruits and vegetables with new eyes.

Even still, micro eco-farmers can concentrate on one specialty crop if that's their passion. They will diversify in other areas, such as their market outlets, and they will use their adaptability to upgrade continually, which keeps them out of the trap that centralized mono-cropping created. Their adaptability makes them nearly immune to either their crop falling out of favor with their customers, or the opposite, "product takeover" where they introduce something that becomes so popular, it is eventually mass-, and cheaply-, produced.

Bob Muth of New Jersey grows mixed vegetables, but specializes in peppers. However, the pepper varieties he sells now are completely different than the ones he grew two decades ago. Bob adds up to five new varieties each year. Working directly with his customers, Bob carefully replaces old varieties with those that outdo them in performance and customer preference. In contrast, megafarms become entrenched with producing one specific crop with machinery investments, fertilizer contracts and harvesting and selling contracts specific to that one crop, and so they can become trapped when the crop falls out of favor or is more cheaply produced elsewhere. But the micro eco-farmer, who is always siphoning in new ideas for products from his/her community, and whose greatest equipment investment may be a hoe and a composter, does not fall into this marketing trap. Bob found a way to continue with his specialty in a gentle manner by keeping one ear to changing customer needs and integrating changes gracefully over the years.

A rancher of larger acreage argued that smaller farms couldn't possibly profit. If they happen to find "one product" that does sell at a high enough price, he insisted, soon others will move in and the market will be saturated. The whole point, once again, is that the successful micro eco-farmers don't play that game. They operate in an area floating above it. In the first place, their products are very hard to duplicate because they are selling that whole orchestrated picture attached to the product, not just the product. They see new opportunities before anyone else does, are extremely adaptable, and do not have all their investments in a single-crop with a single market. They gradually siphon in the new and gradually release the old before anyone else even knows what happened. They are in a perpetual state of renewal and upgrade, just by their nature, and just as in Nature.

Art Biggert and Suzy Cook, owners of the 1.55-acre suburban Ocean Sky Farm, started out as a 75-household community supported agriculture (CSA) farm (see more on CSAs below). "We never had any trouble filling our subscriptions to the CSA," said Art. "People frequently commented on the difference in texture and flavor of our fruits and vegetables that were allowed to ripen on the plant and picked fresh for their table. We encouraged our subscribers to visit the farm when they picked up their produce by offering a U-pick flower choice each week. We planted different flowers all over the farm in order to attract and house beneficial insects as part of our pest management program. We also made all of the fruit choices a U-pick. This saved us a lot of labor and gave our subscribers a reason to pause, reflect and teach their children about where food comes from." In what Art describes as a tough decision, he and Suzy decided to exchange their vegetable beds for perennial medicinal herbs, which earned them even more income in their particular location and was an area of interest to both of them. It went along well with the on-farm body care products they processed.

When emphasizing one particular crop, farmers can also craft a diversity of on-farm products from the crop to appeal to a wider variety of interests. On Blueberry Ridge Farm in Sherwood, Oregon, Gary and Lynn Haase have raised blueberries since 1983. Starting as a fresh blueberry farm, they branched out to include jams, jellies, dried blueberries, canned whole blueberries, oat bran blueberry muffin mix, and blueberry buttermilk pancake and waffle mix. By diversifying their blueberry products, they're diversifying their customer type. Some people prefer to pick their own and make their own pancakes from scratch; others enjoy the work already done, right from the farm. By appealing to a broader range of customer preferences, Gary and Lynn are building their customer base without having to appeal to (and depend on) a single customer type. Each type on its own may not provide a large enough base, but collectively, they add up to a steady stream of customers.

Micro eco-farmers can also change packaging quickly. When Melanie and George Devault of Emmaus, Pennsylvania, found themselves sharing a tomato glut with other vendors at their farmers' market, they turned big boxes of under-priced tomatoes into profitable salsa kits—a recipe, a few tomatoes, and the other herbs and vegetables called for in the recipe. What they sold, then, was the labor that went into researching a homemade dish, and the perfect amount of produce needed, already harvested and sorted and ready for the cook.

Microfarms rise to the challenge of educating the community on how purchasing from a sustainable farm is also rebuilding their earth, and the pricing that requires. A belief still lingers among the mainstream that very cheap food is a money saver and a better deal. When micro eco-farms educate their customers and stand firm to their value on this planet, this converts even more new customers as well.

Micro-farmers may turn their crops into salsas or pie fillings or salad mixes or old-fashioned sourkraut from secret family recipes. In some states, kitchens have to be certified at great expense to the farmer to legally allow on-farm processing. To get around this, farmers make arrangements with local churches or other public facilities that have certified kitchens. Community kitchens specifically for small-scale farmers are becoming available in some communities.

The micro eco-farm's direct "local" community can also expand to mean community of shared interest. Singing Brook Farm creates "flower essences," which are healing remedies made from flowers and are lesser known to the mainstream. While owner Mariam Massaro sells locally, she also sells worldwide direct to her "village" of people who enjoy these one-of-a-kind remedies. Green Hope Farm of Meriden, New Hampshire, also sells its version of flower essences via the internet. The owners correspond directly and regularly with their customers, getting direct customer feedback and building strong relationships.

Direct contact with the community

Microfarms reach out into their community for resources already available, (such as the supply of leaves delivered to Bird Haven Farm described above) and they also rise to the challenge of educating the community on how purchasing from a sustainable farm is also rebuilding their earth, and the pricing that requires. A belief still lingers among the mainstream that very cheap food is a money saver and a better deal.

When micro eco-farms educate their customers and stand firm to their value on this planet, this converts even more new customers as well.

When microfarmers give classes or hold demonstrations they sometimes simply put up a sign that might read something like:

> *No serfdom is created here.*
>
> *No earth is destroyed here.*
>
> *Those who intelligently, slowly and carefully hand select our crops for harvest are either well-respected interns learning ways to earn an independent, healthy living on their own, or are people receiving more than poverty-level wages and live in real homes rather than 25 packed in a mattress-filled shack.*
>
> *Our potatoes have skins so tender they melt in your mouth and our baby squash cannot be held more than two at a time to keep its tender skin perfect. Our peaches are so juicy they have to be slowly and gently taken from the trees and our sweet baby carrots are so brittle and delicate they must be carefully hand dug.*
>
> *All of our produce is selected for quality when biting into, not for mechanical harvest and long-distance shipment, and we value those who harvest such produce and we pay them well, just as we would pay you if you were one of our harvesters.*
>
> *Our farm does not add to the billions of dollars spent each year in ill health, unneeded deaths, toxic waste transport, and environmental destruction. However, our farm does restore health, life, and regeneration of the earth and her biosphere.*
>
> *Therefore, our prices are set as they are.*

To continue direct contact with the local community, micro eco-farmers reach out in a variety of ways. These include an internet presence and newsletter, free publicity from news releases sent to local newspapers, yearly on-farm open houses, and donating produce to community charitable causes. One farmer brought his sweet corn to a community fundraiser and sold fresh, hot cobs dipped in real butter. Local farmers bring free samples direct to chefs, donate a free bouquet to the local library with the farm's name displayed, and add a basket of their bounty as a prize for a contest for local charities. They may even print recipe books of customers' favorite farm cuisine. They add income and keep their community connection alive by offering gardening classes, herbal soap-making classes, school farm tours and cooking classes.

The direct markets for their products already mentioned also keep micro eco-farmers in contact with their customers. Such markets include on-farm stands, road-side stands, U-picks, community supported agriculture (see below), home deliveries, overnight express, direct-to-local restaurants, inns, bed and breakfasts, catering services and juice bars. Micro eco-farmers cater to shared-interest groups such as hand-spinners, home gardeners, hobby chefs, and ethnic communities. Some serve specific spiritual groups. Bob Russell runs a farm on less than one and a half acres in a seaside resort town in Delaware. He offers custom growing of a wide variety of herbs, vegetables and edible flowers to local

This backyard farmer's crops at the farmers' market change throughout the season, from small spring garden starts to larger tomato plants to ripened heirloom tomatoes.

restaurants. He never turns down a chance to give VIP garden tours to chefs, which result in increased sales almost every time.

Community supported agriculture (CSA) farming has grown in popularity in microfarming. The farmer benefits greatly from the direct customer contact the CSA generates. Customers "subscribe" to the farm, or buy "shares" in the farm, and come to feel it is partially their very own farm. They pay in advance for a season's worth of weekly produce, whatever is ripe that week. CSA farmers try to offer a large variety for their customers, at least 15 different items, along with ideas on how to serve the produce. In turn, the farmer is able sell whatever ripens on the farm at whatever time it ripens. A farm newsletter keeps shareholders up to date on spring plantings and the new chicks that arrived, and customers are also known to volunteer work on "their" farm.

CSA cooperatives allow micro eco-farms to pool their advertising resources yet remain unique and diverse, and still get close to 100 percent of the dollar. These cooperatives are a collection of small farms and microfarms that each offer something unique. Different than the old model of the farm cooperative, where several farmers growing the same crop pooled that single crop for wholesale prices, the cooperative CSA's strength is the diversity and uniqueness of each farm, which is still allowed to sell direct to the customers outside the cooperative. A cooperative CSA may offer its customers shares in weekly vegetables, weekly fresh-cut bouquets, weekly on-farm organic yogurt, or shares in the blackberry and grape season. When a local small orchardist joins in, they can also offer organic apples and cider as an option. By taking on more farms with more options, they attract even more customers who may then be inclined to purchase products from the other farms. Each farms' crops don't compete; they sell retail; and each new farm helps the others become more valuable. New Hampshire orchardist, Michael Phillips, author of *The Apple Grower: A Guide to the Organic Orchardist,* sells a portion of his apples in a cooperative CSA, allowing customers of that CSA to choose organic apples and cider as one of their options. Yet, this orchardist is still free to market to on-farm and other customers outside of the cooperative.

Jim Sluyter turning compost at Five Springs Farm (see p. 48).

Another cooperative method of reaching the public is gift baskets sold during the holidays or summer festivals with products from several local farms, and a ticket for the summer "farm tour" where several local farms hold an open house during a weekend in September.

While selling direct is the mainstay of most microfarms which keeps them in touch with their communities, occasionally there is a wholesale niche they fill, at least temporarily. Michael Phillips, the orchardist mentioned above, sells apple products on-farm for those who like to visit the apple mill and walk through the orchards. He sells direct by mail for those who want his type of apples, grown on his farm, in his location. And he also sells wholesale to local natural foods stores. Food co-ops and natural food stores often feature the specific farm where the produce comes from. In arrangements like this, if the farmer has surplus beyond direct sales, this type of "wholesaling as a sideline" can serve to sell remaining apples quickly, as well as build up the farm's reputation and get his name out for those who may then decide to come visit.

Resources already available

The low farm overhead that micro eco-farms often have is another example of taking advantage of "resources already in place." Their already-established homes are most often also their places of business. The net income for micro eco-farmers can be higher than that of businesses outside the home, or agribusinesses that invest in large machinery, chemicals and landmasses. People can earn full time incomes at home. "All in your own backyard and all sold in your local neighborhood," said Mel Bartholomew, inventor of the Square Foot Gardening system.

City microfarmers may have higher property taxes, but those taxes help provide them a home in an environment they prefer to live, with many high-paying restaurants as customers. Their innovation at collecting roof runoff water and building their soil to require minimal water cuts into their city water bills. Sustainable farms of all sizes pride themselves in becoming as independent as possible from importing outside soil amendments. Micro eco-farmers take it a step further, from independence to interdependence, and make mutually beneficial trades with other sustainable farms. Art Biggert and Suzy Cook, of the 1.55-acre Ocean Sky Farm, have raised farm animals along with herbs and other on-farm crafted products. When Art found an organic grain producer, he became a distributor for the grain after querying other farms about their need for a local supply. He gets the grain he needs for his animals at wholesale, and local farmers enjoy coming to his farm to purchase grain, which is now another profitable income source for Art and Suzy.

In conclusion, remaining adaptable while orchestrating a diversity of crops, products and community markets is not really a new way of thinking, but more of a return to the natural order of thinking. The initial spark for a micro eco-farm venture may begin first with a couple of specific vegetables the entrepreneur would love to grow. Or, it may start by seeing an unfilled niche within a local community or shared-interest group. Wherever the aspiring farmer starts, with the market or the crop, one will lead to the other. The crops and the eventual market must come together to initiate the process. From this new center, orchestration begins and more diversity radiates out from this center, building on itself and getting stronger and stronger. Eventually, there will be enough diversity and the farm will often settle into more of an ongoing, improvement phase. The farmer, then, is the conductor of his or her own creation, orchestrating diverse components into a single whole that provides for humanity as it replenishes the earth.

"We're also experimenting with something more," Robert Farr, "The Chile Man," continues from his column he shares with us here, "developing a family farm which is completely integrated with the surrounding community, creating a healthy, rural lifestyle we simply refuse to let go."

CHAPTER SEVEN

How to Begin, How to Learn, How to Continue to Flourish as the Earth is Restored

We are operating full time now. Until last year we were still doing our day jobs, which were done here at home. We started seriously doing farming as a part-time job when we moved back to the home place where Jim was born in the corner upstairs bedroom.

—Nancy Carey, Carey Family Farm, Oregon Cascade foothills.

Returning home to the farm (even if for the first time)

It often starts with the right questions: How do I begin with no money? How do I become a farmer if I'm not already (or learn to farm a new way)? How and where will I market my crops? How do I know what the laws are? How much money will I make? The techniques and laws are different for every bioregion, community, and farmer's personality. But the answers are readily available. Many questions are answered by networking with others and seeing examples of how others start, learn and grow.

How do I begin with no money?

"While Suzy was working on her Master of Science in nursing, we went for a February bike ride which a Seattle Bike Club sponsors every year on an island in the Puget Sound," said Art Biggert of Ocean Sky Farm. "I asked Suzy, 'Wouldn't it be nice to retire on an island like this and have a huge garden?' She smiled and asked me, 'Why wait until we retire?' Then, she cautioned me, whispering, 'Be careful what you wish for.'"

Greenhouse starts at Five Springs Farm (see p. 48).

Many programs on reaching goals and achieving success say to start with a conscious vision. This method is highly recommended. However, The Herbfarm near Seattle, Washington is a small acreage success story visited by thousands yearly from all over the world. The originators, Bill and Zola Zimmerman, admitted no such vision that they were consciously aware of. They seemed to use another popular ingredient to their success formula instead. They had fun. They bought the old farmland to retire on and to have fun in the garden. When a few extra herbs were put out for sale, people bought them. When a few more were put out, more people bought them as well. Now, the farm has its own restaurant, catalog, festivals and classes. Perhaps an ultimate formula for success might be a mixture of these: "Have a vision and have fun."

How do I become a farmer?

"During the time between part-time and full-time farming, we planted 3,200 fruit trees on trellises, put in pasture fences to make better use of the grazing, tore down what was left of the original homestead barn (built in 1893) and put up a smaller barn with a lot less charm, but one that would not fall down on us!" said Nancy Carey. "We are returning the fertility to the soil and cleaning up debris from 30 years of the farm not being used. Our acreage totals 40-plus but we have about six acres in production." The Careys fruit trees are intensively planted and trellised on three of the acres. The rest is in greenhouse/shade house, a few acres for small livestock including chickens, goats, sheep and one llama, two bee yards, and four plots for veggies and flowers. "A lot of the land is unproductive because there are four terraces dropping down to the North Santiam River. We have 2,100 feet of river frontage. The goats and sheep love the side hill areas and the trees and rocks give us some interesting micro 'pockets' to do experimental crops in."

While some aspiring farmers work into full-time farming gradually, others need to jump in full speed the first year. "My very first year showed me that, given enough customers, I can earn a decent living," said Baruch Bashan, the software programmer who started Gaia Growers Farm on one-half acre of leased land. "Since losing that first lease, I've been scrambling to find two – five acres in any of the counties in and surrounding Portland. My challenges, at least to start, are to find enough local land/water to lease and to gain enough customers (translate: afford enough advertising). I need to make enough the very first year, every year, so I need to go for 50 – 100 customers right off the bat."

Vegetable starts in recycled yogurt cups at Island Meadow Farm.

Organic produce display at a natural foods market in San Francisco, with point-of-purchase signage detailing grower information for each product. The signs include location as well as production practices (IPM, certified organic, biodynamic, etc.) for each grower. Retail stores such as natural grocery stores and restaurants that support and showcase local growers are becoming more and more popular.

Art Biggert and Suzy Cook started out gradually, but came to a twist in the road. "The greatest obstacle for us to overcome," he continued, "was our fear of poverty. It is difficult to let go of the familiar. By the grace of God, the familiar let go of me and I got laid off. Our fear of poverty was remedied by a six-month severance package and unemployment benefits. It even started in the early spring (1993)." So the couple plunged in full time.

"We decided to apply 'See one; do one; and teach one' (teach by example) to our farming process," he continued. "So our mission statement reflected our intent from the beginning. The mission of Ocean Sky Farm is to observe, apply and teach sustainable agriculture in a suburban setting. All of our sales were to be direct to the customer to provide a teaching opportunity. We'd utilize a CSA, farmers' markets and a mail order service on the internet."

Flowers at Fertile Crescent Farm (see p. 116).

How do I learn to farm a new way?

As far as the hundreds and hundreds of possible crops you can grow and the animals you can raise, networking and self-study have been the teachers of most micro eco-farmers. Regardless of the chosen assortment, you will weave a brand new pattern like no other. Art described he and Suzy's situation before his lay-off. "So, we bought the farm and started to research gardening," said Art. "Since the farm we bought was overgrown with blackberries and Scotch broom we had lots of time to read and learn about fencing while the milking goat cleared the fields. We learned a lot about how water and light moved across our land. We started to plan and build the infrastructure for a sustainable growing operation. We gained an appreciation for the way the proper livestock can save labor."

Micro eco-farms don't often do things the conventional way, although the conventional way can be a place to start. The organic and conventional methods on all crop and animal raising are available in the form of local, on-farm demonstrations and internships, videos, books, periodicals, workshops, associations, and extension services for your particular region. They are also out there in the form of your local garden club and the neighbor who has gardened in your area for the last 30 years. As well, sustainable production is improving, and on-line networking is firing across the internet as you read. Laws on certifications, food processing and home businesses are unique to each locality and change often. If you're open to online networking, there is hardly a question via the e-list communication that won't be answered for free:

"Hi, everyone. Does anyone know how to get your bank to cash checks made out to your farm's name?" —Tom

"Try registering your farm with your county to get a 'DBA' number." —Jack

"Thanks, Jack, but what's a DBA number?" —Tom

"That means "Doing Business As." Lets them know there's a business in your area under that name and costs about $25." —Jack

"Thanks, Jack, where would I look for the phone number?" —Tom

"I found mine under 'Business Licensing,' but I have the phone book for your county, also. Here's the number you should call…"

"Oh, and Tom, I found that I had better luck switching banks." —Sue

Through sustainable gardening and farming publications, gatherings and online networking, microfarmers swap stories of painting a few walnuts red and dropping them into their strawberry patch before ripening time. Birds check out the walnuts to find them inedible, then by the time the strawberries have actually ripened, the birds have already lost interest in that area as a food source. They tell each other how they wipe out all of their livestock's parasites including grubs, ticks, ringworm, lice, flies, mange and fleas, with a simple addition of the nutritional diatomaceous earth to their food supply and animals' coats. They tell where they found handcrafted covered wagons designed for miniature horses, where they found a source for the very rare miniature Guinea Jersey dairy cow. They explain how they built a lighter weight chicken tiller or a more efficient indoor worm bin from a utility sink. The final responsibility, to check out laws and methods, lies in the hands of the farmer, but there are many hands out there ready to extend help.

Sylvia and Walter Ehrhardt who created the famous five-acre Ehrhardt Organic Farm in Knoxville, Maryland, found they were very glad they didn't take expert advice when starting out. Back then, the textbooks and county extension agents warned them that organic fruits would be very hard to grow. They proceeded anyway, planting a half-acre of raspberries the first year, then continued forward with a half-acre of strawberries and an acre of blackberries. Their farm became so successful it eventually became known nationally and internationally.

"Our farm has become a commercial venture as well as a demonstration and educational center," Sylvia said. "We do workshops, seminars, and lectures on-farm and off-farm and give tours for organizations and groups. We began an intern program for those who wanted to learn to grow food organically. Each year we have a few interns who come from the states as well as overseas. These interns learn by hands-on experience as well as by reading and discussing materials. We also belong to the national and international 'workers on organic farms' program and have people come to the farm for short periods from many places."

Under the Resources and Networking section, you will be led to networking groups, associations, forums, workshops, periodicals, videos, books, demonstrations and internships.

How do I know what the laws are?

According to the book, *Sell What You Sow! The Grower's Guide to Successful Produce Marketing,* "Direct marketers who sell to consumers from roadside stands, U-pick fields, community supported agriculture (CSA), or certain certified farmers' markets may be exempt from sizing, standard pack and certain container and labeling requirements. However, nearly all marketing activities are affected by various federal, state county and city ordinances, regulations or rules. In addition to taxes, insurance liability and incorporation issues, farmers need to review labor laws, health department requirements, land-zoning or land-use rules and the permits that go along with them."

The book advises to check with various local state and federal authorities before trying to market any kind of food item: "Check with appropriate state and local officials before you start so that there are no unpleasant surprises down the road. You may be surprised at how easy it is to meet requirements, especially if you're not selling processed food. The regulatory people often will give you valuable, free advice on many aspects of your operation."

Online websites and discussion groups are another place to find information about the rules and regulations that may apply in your area. The internet has made it much easier for private citizens to remain on top of regulations and laws in this new century. Small business was once slowed down by having to weed through manuals and books just to find which laws directly pertain to business owners in their location; yet in this new era of online resources, laws and regulations no longer have to be the discouraging pain and misery that they used to be. Many of these online resources are listed in the Resources and Networking section.

How much money can I expect to make? How will I sell what I grow?

As far as what micro eco-farms earn, that has been discussed somewhat throughout the book, but the bottom line is that they earn whatever they want to earn depending on what type of lifestyle they choose and their value system. They may start out whipping up formulas in the kitchen from their herb garden and selling them at roadside stands, as the originators of Burt's Bees products did, then eventually become a million-dollar industry. They may land a small-farm lifestyle so complete and fulfilling that a comfortable yearly salary brings them all they could ask for. The definition of "full-time salary" often fits that of a grocery clerk, copyeditor or an engineer, although those full-time jobs aren't often enough for a family of five with the other spouse not working also. Yet "real farms" are often not considered "real" unless they earn the income for both adults and all their children without the other taking a second job. Micro eco-farms have fit that category of "real," allowing families to all work together for the one salary they earn from the farm. Yet they have also filled the needs for spouses wanting to earn a full- or part-time income along with their partner who has another full- or part-time job.

Sell What You Sow! cites a study done by the Cornell University Farming Alternatives Program citing "marketing skills" as the most important factor for success in alternative farm enterprises. "Marketing crops is an artful balance between producing what customers want to buy and producing what you love to grow," says the book's author, Eric Gibson. "Unlike the traditional commodity-crop farmers who just dumped their products at a produce terminal or grain elevator, the successful, small-scale farmers give a lot of quality attention to marketing. The have a keen ear to what buyers want and seek constant customer feedback. Yet it's also true that if you grow what you love to grow, your enthusiasm comes through and customers pick up on this. Both are important: marketing skills, as well as having fun at what

Rt.: "The Chile Man," Robert Farr, at an Open House. Above: yarrow and strawberries at The Chile Man Farm (see p. 26).

you're doing!" *Sell What You Sow!* can be found under "New World Publishing" in the Resources section, where you will find many other excellent small-farm marketing resources as well.

Stories unfolding on an earth eager to bloom again

Baruch Bashan secured the land he needed in time, and he since found a complementary passion for his farming, the breeding of open-pollinating seeds.

Sylvia and Walter Ehrhardt now spread their wisdom across the world. "My husband and I were asked to do organic projects," Sylvia said, "first in Pine Ridge South Dakota at the Lakota Indian Reservation, then in Russia. These projects took us away from the farm for a few weeks to a month at a time. Having my daughter, Beth, here allowed us to leave. As we really enjoyed teaching and developing organic agriculture and gardening programs we realized that this was probably the way we would spend much of our time if the farm were managed by someone else. Beth took over the management of the farm, and we spent more time overseas developing programs. Beth now officially owns and operates Ehrhardt Organic Farm."

It's been said that the greatest scientist of all is the earth. She offers millions of years of experimentation just waiting for us to observe closely enough to see the answers she invites us to tap into. One might call it the science of eternity. "Our clinical experiences have helped us learn to live in the present. We each have had encouragement from several terminally ill people to seize the day. We have been convinced that now is our eternity," said Art Biggert of Ocean Sky Farm.

Micro eco-farmers seem to download and demonstrate eternity, undaunted by the appearance of setbacks or others telling them: "It can't be done, there's no scientific evidence for it." They don't scare easily. After all, eternity has nothing to fear.

"In the past four-and-a-half years, we've seen our holistic farming practices dramatically increase the bird population," Robert Farr, "the Chile Man" shared from his column in *American Farmland Trust Magazine.* "A glance out the window offers bluebirds, sparrows, juncos, downy-headed woodpeckers, American finches, red-tailed hawks, red-winged blackbirds, hummingbirds, chickadees, and the occasional blue heron or pileated woodpecker.

"Ultimately," he continued, "Our mutual goal must be to re-create a sustainable economy, where people formed lasting relationships with other people, and real community came from those who lived around us." The small farm—ours and yours—holds an important place in that new world."

Once again, there is a change among those who farm in this century. Are you one of them?

"I have loved the earth all my life," said Mariam Massaro of Singing Brook Farm: " She heals, excites, inspires and brings me my loves, my work and my connection to God—all through the reflection of her beauty: the ocean, the trees, the rivers, the sun and the moon."

EPILOGUE

Growing Children's Souls: Families and Children on the Farm

The kids can't come to the phone right now, Kaleb, Elise is down in the garden eating radishes and lettuce, Jereme's up in his tree fort eating cherries.

—Telephone conversation from Island Meadow Farm

On Bryant Blueberry Farm and Garden in Washington State, owners and husband and wife, Jamie and Leslie Bryant, live on five acres and grow blueberries and huckleberries on a ten acre peat bog across the street. They offer U-pick and farm picked fresh berries, sell frozen berries during the off season, as well as the actual blueberry and huckleberry plants. As an interesting sideline, they operate a dog boarding facility on their land. "It is a separate business," Leslie said, "But blueberry customers have been interested in boarding their dogs here, perhaps because they have had a good experience on our farm. The dog boarding is a natural environment, with lots of room for the dogs to run around instead of being confined to kennel-cages." The Bryants open their farm to the public during picking season and they especially welcome children. Kids not only enjoy picking the berries, the Bryants offer a playground with ducks, geese and twin cashmere goats to watch and pet. Picnic tables are available, and they invite families to make a day of enjoying their farm's scenery.

Not all farms benefit from bringing customers' children to their property, nor will all farmers be parents. However, when children are involved, they often benefit in deep subtle ways, and so does the farm's bottom line.

Growing up on the farm, my two children, Elise and Jereme, ate fresh fruits and vegetables as a form of entertainment. They ripped husks from corn and ate the sweet kernels raw from the cob; dug radishes and baby carrots and ate them right out of the ground; picked yellow pear tomatoes and lettuce straight from the garden and ate them without any dressing. The only problem was convincing them that sometimes humans tend to hold off eating, bring the food from the garden into a square building called a house, put it in hot water, then take it out of the hot water, put it on a round thing called a plate, and eat it with something called a fork. That way we parents can feel we've provided a "hot meal." Thank goodness they never asked me: "why?" I wouldn't have had an answer. Most would reason it is because a family sitting around a table is an asset to family bonding. This is very valid, but my kids were far too inventive even for that reply. Had I suggested it, they'd have invited us all to join them down in the garden and up in the tree fort for family bonding.

Eventually, my daughter started bringing baby cukes into the house to make homemade pickles and brought ingredients inside for mixed salads with her favorite organic French dressing. Spaghetti cooked with fresh garden heirloom tomatoes, basil and oregano from the garden was beyond compare. My son learned to bake great apple pies from our orchard's antique apples. Yet he continued to ask if we could simply leave out a few green beans for him to eat raw on his plate while the rest of us ate them steamed. Although I checked thoroughly, nothing in any parenting books told me that honoring his request would destroy him, so I happily obliged. He freely offered his assistance on the farm, which tended to draw out in others around him the desire to offer their assistance to him.

Family meals around the table really were times of great enjoyment that reinforced family unity. As the outdoors came indoors to nourish us, we also felt a unity with the greater universe. The kids used to love to have meals with as many ingredients as possible coming right from their own property.

Micro eco-farming allows children back onto the farm again, in a real, participatory way

The taste of mineral-rich fruits and vegetables and the taste of sun- and vine-ripened freshness surely play a part in why kids eat healthily so eagerly when their food comes right out of the garden. But there seems to be more to it. When they see it grown and take part in the food's nurturing or harvesting, it's as though a deeper connection to "Something" emerges again and again.

Nancy and Michael Phillips, with their daughter, Gracie, are owners and creators of Heartsong Farm, a unique, mutually beneficial combination of sustainably grown, rare apple varieties, medicinal herbs, and farm crafted herbal products produced on about four acres in Groveton, New Hampshire. Their products are very unique, in that they feel it is not only just the species of herb, but how and where it was grown and processed, that adds to its healing ability. "There's no turning back once you understand how much of a difference high-quality herbs make in a holistic healing approach," Nancy said. "Good dried herbs look, smell, and taste very similar to the living plants out in the field and garden. Herbal products made in small batches by community herbalists feature that same superior quality."

The Phillips are living in a world that does not exist in children's textbooks nor on TV, and as part of their farm's profitable offerings, they give herb and nature camps and classes for both adults and children. Magical things are happening on micro eco-farms everywhere, and on Heartsong Farm, where else could one experience first hand the kinship with the earth as the Phillips live on a daily basis? "Plants are ancient beings," said Nancy. "We have evolved on this planet with them, and accordingly our bodies know how to work synergistically with the plant constituents as well as subtler energies. Our spirits are nourished as well in our relationship with plants. Sitting with these

Child in kale at Salt Creek Farm (see p. 29). Photo taken by CSA subscriber Martin Hutten.

ancient ones can sometimes be just as powerful as ingesting the herb, if not more so."

To absorb this type of knowledge on a holistic level, children often respond well to being immersed in it in a hands-on, experiential way. "Many children grow up removed from the natural world," Nancy said. "More of their time is spent in buildings and vehicles rather than out in the woods and meadows. Yet what delight they show when they finally get their hands in the soil or wade out in the brook! I simply love having 'Nature & Spirit' camps for children here at our farm each summer. We go out and learn about individual plants, feeling and smelling each herb. The kids get to pick their own salad vegetables from the garden, plus some wild edibles in the field. You can't find a better way for people to learn where their food and medicine comes from than this. Kids need to know that nature is indeed an option in their lives."

Yet inviting children to one's farm is not to be taken too lightly. I have seen well-intentioned plans for children's programs stop in their paths when the adults involved became disillusioned, shocked that some children today are so removed from the natural world and do not immediately show reverence to it when exposed at first. "Having young children to the farm for a week's worth of activities does take a good deal of planning," Nancy said. "Yet once the preparation time is over and the kids arrive, we have just as much fun as they do." Michael agrees that besides the customer incentive and income involved, there's just plain inner rewards for the farm family when children are active farm customers, including making sure adults don't allow the business of the farm to dictate their lives to the point of forgetting how to have fun.

"I think the nature camp idea fits right in with doing things with your own children," Michael said. "Gracie loves having other kids here. We put on nature skits in the teepee—which might never get set up if I let the farm rule the order of my day. Nancy helps the kids create an herbal first aid kit with safe remedies the children make themselves, like Peppermint Tummy Potion. The process tends towards the hilarious, which can't always be said when adults focus solely on making a living." Michael went on to explain how spontaneous joy can happen even beyond the activities planned. "The kids come out to the orchard as part of a scavenger hunt, and somehow we get to following the trail of a fox, finding scat, and even playing at being fox kits scampering on the big boulder ourselves. I guess what ol' Daddy Fox is trying to say is that there's more rewards than a farmer might think in this kind of venture."

Whether your own children are involved with your farm, customers' children are invited to your farm, or your farm revolves around children (such as a school farm), if given enough exposure and prompting through socially conditioned resistance, children seem to remember how to live from a Garden.

Once back in the Garden, one is not separate from health; it's just part of the whole picture. Far away at the University of California, a professor has announced results of a study proving that organically grown crops contain up to 58 percent more polyphenolics, a healthy antioxidant compound. Young children on the farm may grow up to be similar-type professors, conducting research to help build a better world. They may grow up to be the next generation of earth stewards.

But for now, once they get enough of the taste and rhythms of nature, something inside tells them that this is where they're supposed to be, regardless of studies that prove their case.

From what started as a seed potato farm selling out of a pickup truck, Ronniger's Potato Farm, owned by the Ronniger family, now sells heirloom seed potatoes across the country. Their catalog listings sound like potatoes from a colorful storybook, with such varieties as "Maroon Bells" which has dark maroon flesh and skin, and "Viking Purple," a purple-skinned potato splashed with pinkish red, and snowy white flesh. The family stewards about 100 acres of forest and natural wetlands and farms 20 acres within this paradise, which is at the foothills of the Purcell mountain range in northeastern Idaho. As a sideline, they raise Haflingers, a compact draft horse breed known for its loveable personality. Both the Ronniger children and customers' children have had the privilege of experiencing the farm. "They develop sense and sensibility about what to do and what not to do," said David Ronniger about watching children on the farm. "It allows them to know where their food comes from, to know life from a seed, a flower to smell that came from the seed they planted, animal husbandry and birth. To see a birth is almost a miracle to any child. To see a horse born is something no child will ever forget."

Again, this is not to suggest there are not many children who've moved away from their innate knowing and don't carry a lot of anger and adherence to commercial values. It can sometimes take a long time, and a long, supervised program, to return such children to that voice within themselves that's connected to their natural world. Many children have lessened their ability to be in tune with, and interact with, another living thing. They "interact" with their computers, but so far, the computers don't find it tempting to react back by nuzzling them in the ribs for more attention. Children may have to relearn the art of thinking about the thoughts and feelings and possible reactions from another they can't control as they do a machine. Farm animals may need protection from visitors' children left alone with them. Even your cherry tomatoes may need protection from kids far removed from the natural world. But time and time again, if given enough faith and supervision,

Author's daughter, Elise, surrounded by children eager to pet her horse.

children caressed by nature's wonders have been healed in numerous ways.

A new farm product: reviving the hearts of children

Ned and Jackie Bowen, owners of Wind Hollow Ranch in New York, offer horse riding and caretaking to young adults with special needs. They have described horses as animals who can heal emotional wounds and are therapeutic for physical problems as well. "Equitherapy" is becoming a more common term and has been reported to assist in developing equilibrium reactions, normalizing muscle tone, developing better head and trunk control, coordination and spatial orientation.

Acres USA recently reported on a five-year experiment on healthy eating at Appleton Central Alternative School in Wisconsin. Concerned about high dropout rates, expulsions, drug abuse and other negative behavior, the school eliminated all junk food for five years and replaced it with fresh juices, bottled water and a salad bar. The negative behavior patterns dropped to almost zero. There is just something about the natural world that heals.

Author's son, Jereme, in tire swing under cherry tree.

My daughter, Elise, noticed her pony lying down sleeping on a sunny day, and cuddled next to her to take a nap. Elise was, by now, so tuned in to horses that in a split second she had the instincts and motor skills to leap up to safety, even during a nap, if her pony became startled. For another child not as tuned it yet, this act could have been dangerous. And yet it was so valuable an experience. Elise announced afterwards that during that experience she felt completely at one with the whole world.

My son, Jereme, rescued birds since he was old enough to go outside alone. When Jereme was around 5-years-old, we had another visiting 5-year-old boy I'll call Mike, known to be angry, the type who applauded and laughed when hearing descriptions of other people's pain. Even at a young age, Jereme could read the character in another with lightening speed. When Jereme and I saw Mike heading for our down-covered ducklings, our eyes met in mutual, "He'll kill them!" and we rushed to the scene, trying to be cool about it all, knowing this boy would feed off any special attention his behavior drew to himself. Ever so casually on the outside, with hearts pounding on the inside, ready to leap to the ducklings' rescue in a second's notice, we watched his every move. He looked closely at them. Then kneeled down. Something about him changed. Jereme detected this and relaxed. I relaxed (knowing I could trust Jereme). Mike's fists softened. His expression changed to awe. And gently, he slowly lowered his right hand down to touch their yellow down. Some instinct had told him not to move too quickly to keep from scaring them. Something whispered to him he'd be rewarded for gentleness. If ever I saw strength in a young man it was now within Mike: One so entrenched in the idea that cruelty was a sign of strength, choosing on his own with no outside prompting, at least just for that moment, to go in the opposite direction.

I wondered what would have happened to this boy had he had continued experiences such as these. For this was the type of choice and strength, listening to an inner knowing instead of an outside crowd, I witnessed from my own children on a daily basis that carried over into their adulthoods. The farm is loaded with opportunity to draw out the best in children. And this alone can be a "crop" that farmers harvest, even for revenue.

Our pumpkin patch project, for example, was a yearly highlight on the farm. It started with Elise and Jereme planting big, pointed seeds, which in turn led to a lively, bustling, one-day autumn festival where members of our community came to Island Meadow Farm for a day of old-fashioned harvest fun.

The children planted seeds of many pumpkin varieties, from big, future-giant pumpkins, to heirloom "Cinderella" pumpkins, to the delightful, munchkin-sized ones, along with miniature ornamental corn and a collection of ornamental gourds. They nurtured the seedlings, transplanted them, and continually cared for the pumpkin patch garden.

Before the big day arrived, fliers had been passed out at the local elementary school for our chosen "Autumn Festival" day in October. The children worked with focused anticipation to prepare. A board was laid across two bails of hay for a table. Two buckets were turned upside down for stools, and handfuls of change were divided into a metal cupcake pan. The "cashiers' counter" was now set up. The ponies and pigmy goats were groomed and haltered, waiting to entertain customers when they arrived.

Kinship with nature/ faith in the unseen

> *In a society that still promotes instant gratification, burying a whitish fleck six months before the arrival of any fruits, and daily pouring water on nothing but soil calls forth a beautiful remembrance of faith in an "unseen something" greater than ourselves.*

In a society that still promotes instant gratification, burying a whitish fleck six months before the arrival of any fruits, and daily pouring water on nothing but soil calls forth a beautiful remembrance of faith in an "unseen something" greater than ourselves. The reward comes in silent joy when—because of our faith and trust—a previously non-existent green plant pushes up through the soil days later. New leaves form; a vine lengthens, thickens and twists across the garden; smooth tender green bulbs swell from where there was once a yellow flower, inflating larger and larger, transforming to yellow, and then at last to the anticipated orange we all recognize—all from these two humans having faith in a hard, button-sized seed. Partnering with nature this way draws out patience, reverence, and respect—not only respect *for* nature but also the sense of being respected *by* nature, who trusted these human stewards enough to co-create new life under their management. It is a good feeling to sense that something as powerful and complex as nature sees us as an ally.

Coming full circle

Although I describe this project as "starting with planting the seeds," my kids actually witnessed the full circle of life as a cycle, rather than the more fragmented notion of a beginning and an end. The circulation went from pony- and goat-manure compost made from manure, to planting, to harvesting, to sending out the fruits of their labor to the world, to picking up manure from the pets who entertained customers, and back to compost again.

Attaching nature to human society

By working with nature and then having the results attached to human celebration, my kids were allowed to experience a connection between human society and the earth, a synergy between the two. To feel something grand between humankind and the planet that neither human nor the natural world could have created on its own helped my children internalize something far different than the desire to dominate or escape from nature that still infilters much of their media-driven world. At best, modern society suggests that the earth is something weak that must be rescued by humans, when in fact the children were able to experience that nature has a power beyond humanity, and will empower any human who wants to join it.

Blending "process" with a tangible goal

Sometimes, children's quest to reach goals too quickly can erase the beauty of the process that gets them there. To counter that, a segment of society suggests children maintain no goals at all, that they float along from anywhere to anywhere so they learn to be in the moment and in tune with the now.

I found that by creating a garden with a tangible future goal of a harvest festival blended the two partners of creation—the in-the-now journey of process and the attainment of the tangible goal—quite nicely. The desired goal channeled the energy and motivation; yet the children had to slow down. You simply cannot force nature to bring October any faster than it's going to get here; you have no choice but to go at a slower pace.

Being a contributor to their community

Children and families in our community wanted to purchase pumpkins and fall ornaments each October. That was a given. Elise and Jereme provided something valuable to their community, allowing them to experience the joy of being a real contributor to their "village." They were not selling wilted dandelion boquets or crayon drawings (as sentimentally precious as those are). They produced and sold an authentic

Young girl selling produce at Sacramento Central Certified Farmers' Market.

product, and the people coming to their Autumn Festival were quite serious about their purchases, looking over each pumpkin, deciding whether it would fit on their porch well or work with the face carving they had in mind.

Being supported by the village

As coins were traded for pumpkins, gourds and ornamental corn, my children felt the win-win situation of village support. They never knew who would show up that day, as one-by-one they pulled back the husks of the ornamental corn to reveal the jewel-like color combinations hidden inside. Would the village really be here? As the kids set the price sign on the pile of corn, and wrote up price posters for different sizes of pumpkins, a quietness came over them, almost as though they were listening for footsteps for their fellow humans to arrive.

The fliers had been sent home a week before in numbers high enough to (one would think) ensure at least a small turnout. And turn out they did, leaving with the treasures and fun of finding their own pumpkins right from the patch and petting the ponies and pygmy goats. For this, my children received payment for some of their own desires that needed financial backing, sensing the real life meaning of "cash flow." Moneywise, this project was aimed at earning for my kids, but our farm became better known from this one project and much word-of-mouth promotion for its other projects and products was generated.

Potato Power! Kids enjoying newly-picked spuds at Salt Creek Farm (see p. 29).

Children who visited the farm, whether from the Harvest Festival or at other times, often left with a little more of themselves than they had when they arrived. A junior high boy I'll call Sam, with a permanently-set frown, was brought to the farm by his parents. I was made aware he had just been beat up at school where he'd had his lunch stolen regularly, and was now suspended for fighting back. He came up to me and asked if he could go into the garden where my two miniature horses grazed.

"Sure," I said, and then repeated the usual rules, "Just remember if you want them to come up to you, you have to move slowly and offer them affection. No fast hand movements around their faces; never chase horses." In that moment, Sam was reminded of another way of being, one quite the opposite of what had happened at school: putting yourself in another's place, and offering love, are powerful tools, more powerful than beating up the horses and stealing their lunch. I watched from a distance as the two little horses trotted eagerly over to Sam. They sensed they were safe near him, and he sensed their trust in him also: an empowering feeling.

In summary, micro eco-farms are nurturing the next crop of humans in many ways, and the world is harvesting the hidden "products" their farms provide these children. In some cases, these products directly generate the farm's revenue as it is with the therapeutic horse farms. Other times, children are added as a sideline bonus. After one micro-farmer invited school children to tour her farm for $2 a head, the children went home to tell their parents they wanted to return for other paid farm projects and products. Including children in the farm takes well-thought out plans, supervision and patience, but the rewards can be many.

On eight-acre organic Salt Creek Farm in Port Angeles, Washington, parents are encouraged to bring their children to the farm and help in any way they can. The farm has operated as a successful community supported agriculture (CSA) farm since before the last century turned. Along with the usual variety of garden-grown produce, Salt Creek offers an especially colorful assortment of rare potatoes, such as "Ozette," one of the farm owners' favorites. "It is a fingerling potato," said Doug Hendrickson, who owns the farm along with his wife. "It's named after a Native American Village on the coast of the Olympic Peninsula. Its yellow tubers are from two to eight inches long with thin, yellow skin. Flesh is creamy yellow and flaky when baked. It's said to have been brought from Peru by Spanish explorers to the Makah tribe living near Neah Bay, Washington State." Allowing children to be part of this kind of history has been quite enjoyable for Doug. "Digging spuds is a blast with kids," he said. "It kind of reminds me of an Easter egg hunt." With children, like the seed planted and watered by the gardener, one could ask, do we grow something new, or do we release something already waiting inside? Perhaps as we give them blueberries to pick and cherry trees to climb we are, indeed, nurturing what's already waiting to unfold, while always aiming to sculpt it closer and closer towards heaven.

Fond memories of May on Island Meadow Farm

I loved creating momentary historic and nostalgic experiences on our farm for visitors. And today, we were about to celebrate May Day the old-fashioned country way.

In the early hours on May 1st, I scurried across the dew-dappled pasture with our pony to prepare for the day. A bouquet of lilacs had just been picked and would add to the bounty of flowers about to arrive on the farm. This valley where my farm lies has been farmed since the 1800s, and it's the home of an original white-steepled, one-room schoolhouse still standing, alone and silent, just as it did a hundred years ago for local farm children.

A group of elementary students from a classroom of eight through 12-year-olds were bused to my farm, and soon that little white building would rumble with school children again. It had doubled as the kids' school and local meeting place for farming families long ago, where funds would be raised for the school by auctioning off the farm women's pies and other baked goodies. And today, it awaited children donned in knickers and suspenders, pinafores and petticoats. They were engaged in a full-fledged historic re-enactment of life 100 years ago.

After the bus dropped the children off at the farm and pulled away, quietude ascended as the children gathered around our strawberry roan pony to watch their teacher harness her to a cart. Jeff, the farm's gander, honked at the excitement. The children enjoyed his antics as he vied to be the center of attention, and, funny, those children who also lived to be the center of attention scolded him for that. Then they settled down themselves, now that they'd had words with that part of themselves waiting to grow up.

Then, a journey that rarely occurs in our modern time began. We walked amidst birdsong and sunshine up the country road to school, taking turns in the pony cart.

This is when it first started to happen: the revival of yesterday's nature-filled country soul that appears when enough is actually re-enacted.

> *I hoped a bridge within them had been built to the feeling of what life is like when it intermingles with nature and that inner voice they can hear at last.*

Canary yellow goldfinches became more visible. The whirling buzz of hummingbirds was astoundingly audible, where on another day it would have gone unheard. Young hands touched dew clinging to wild rose shrubs, and bright eyes searched for possible wildflowers to add to the garden bouquets they loaded into the cart's wagon. Older girls held the hands of younger students. Boys rushed to the rescue when the pony's harness looked a little loose. And glimmering in the morning sun, tin lunch pails held cheese chunks and apples that the kids demanded over commercial junk foods… just for today.

When we arrived at the schoolhouse, kids begged to pull the thick cord that sounds the school bell. As the clear tone rang its nonintrusive song through the surrounding farms and woods, children scrambled to line up to march inside. One could almost feel the little schoolhouse breathing again, as children filed one by one through its heavy red door.

Surprisingly, the feeling faded quickly when reporters and photographers arrived with high-tech equipment, parking their cars nearby. After they left, we requested that no other "horseless carriages" come near the schoolhouse and that visitors in modern garb keep their distance. We were left alone with yesterday again.

The day, enjoyed so much, went by too fast... but only because it slowed down so well. Memorized poems were recited. An elder school marm returned to this schoolhouse that once was her second home, bringing with her antique books and boots and other treasures—including true stories—she'd kept from bygone days. The stories included one about a team of escaped horses and plow that flashed by the schoolhouse window one morning during lessons. Soon, their shouting farmer followed, only to have the team flash by again in the other direction, with the farmer rerouted also, still shouting. From this elder, we also heard of the original hot lunches. After she'd light the wood-

Petting a horse that grazed across from the schoolyard.

burning stove, children put their glass jars of last night's leftovers in a large pot of water on the stove. By lunchtime, hot meals were served.

As the day went on, May baskets were crafted and filled with flowers, then delivered to nearby rural homes. Children saved their apple cores for the farm animals: the ponies and of course, Jeff and the girl geese. They played red rover, kick-the-can, marbles and jump rope... all activities so far from their usual media-driven world that they could allow a part of themselves to emerge that is too often weighted down by their post-modern reality. Yes, here, far away from that reality, it was okay to hear an inner voice that has no need to conform to outward commercial demands.

And it was okay to dance, also—even for boys. The May Pole was lifted, its satiny ribbons streaming in the breeze. Each was held by a child. Then, encircling the pole, boys went under, girls went over, in the classic May Pole dance until a tapestry of spring colors covered the pole.

Amidst all the activity, a quiet stroll had me looking closely at individual children while they were off-guard. Inhaling the scent of a friend's bouquet, enticing a horse on the neighboring farm with a handful of grass, they were experiencing a world that surrounds them still, but had too easily faded from their childhoods. In these moments, I hoped a bridge within them had been built to the feeling of what life

Young hands touched dew clinging to wild rose shrubs, and bright eyes searched for possible wildflowers to add to the garden bouquets they loaded into the cart's wagon.

We walked amidst birdsong and sunshine up the country road to school, taking turns in the pony cart.

is like—even day-to-day schooling—when it intermingles with nature and that inner voice they can hear at last. And may this bridge be one that stays firm within them forever.

Perhaps building those bridges is the reason those of us, in our many different ways, keep connections to nature alive, where we can leave behind the distractions and demands that seem to mask the source of our real home and keep us looking outwards for a place that never seems to arrive. The flock of geese approached the kids' offering of apple cores. Jeff came first, of course. I watched the kids' expressions closely for what I knew would happen next. Jeff stood at the head of the flock, chest out, a feathered incarnate of the word "macho," grabbed the first apple core... and then did something unexpected. Slowly, keeping one eye on the kids, he twisted his long neck and gave the apple to the other geese before himself.

And he continued for every apple core offered. He took it, turned his head, and fed the others first. The kids got quiet. They looked at Jeff differently, now, not quite sure if geese are human, or humans are geese. There was some universal truth for all living things that Jeff portrayed so well. Far beneath all wrong turns and bad tries and attempts to over-control, there lay inside, even in Jeff, a reason for it all that originated in love. Perhaps there was nothing really so bad out there in the world that needed to be destroyed, but instead, its home, its origin, just needed to be returned to.

And many returns to an inner home were revealed on this day. Only 24 hours ago, the child I now looked at did not know a lilac from a tulip. She now delicately arranged them both in her May basket with the precious care one would give the most divinely created artistry. Here, she was tuning in to the essences of little girls that walked this land long ago, as well as with the essence of nature—of something greater than herself. Here, she was not alone, and could, therefore, be alone with her flowers to find a stillness inside where the authentic heart of a child resides, and where the seeds of a future adult's wisdom and peace of mind lie waiting... always there, not just in history, nostalgia, or yesterday, but on whatever road we take to remember eternity.

New Possibilities: Additional Farm Themes

Following is a sampler of 25 unique micro eco-farm themes not described in detail in the main part of the book, all taken from actual, working farms and farmers I've seen. They might serve as inspiration for a theme for your farm, or you might want to use any combination of several of them as subthemes.

- *Layered farm.* On six acres in a very sunny, hot location, this farm grows large chestnut and walnut trees at 60-foot centers on three and a half of the acres. Smaller filberts are interplanted in the dappled sunlight. Peach, plum and pear trees and grapes edge the grove where they receive more sun. The farmer's own bees pollinate and his peafowl patrol for insects. The remaining two and a half acres are for seedlings and grafted whips. This two-story grove was producing $12,000 from nuts alone when I checked in near the end of the last century. The well-nurtured young trees are highly valued and are sold out two years ahead, which gave the farmer additional income of $20,000 just for the young trees.
- *U-milk dairy* blended with a community supported agriculture (CSA) concept and rare-breed dairy cows. Milk is sold through a CSA agreement where customers purchase "shares" in the cows (the ones I researched created a contract with an attorney) allowing customers to obtain grass-fed milk and use it raw if they choose, and milking their own cows as an option. While raw milk may currently not be legal to sell, citizens are allowed to own their own

cows and do as they please with their own milk. Some of the more rare milking cow breeds are better adapted to hand milking and producing from pasture, and the calves can find homes with other breeders of rare cattle.

- *Children's first experience with horses.* A miniature horse ranch created a level of horse interaction for the really young set. Rather than saddling and riding, miniature horses are brushed, lead around, loved, and haltered to take children on cart rides.
- *School farms.* These provide elementary schools with nutritious lunches and whole learning experiences where science, math, history, health, reading and writing all come together on the school farm. Studies and observances are mounting that when kids grow their own food, they'll choose healthier food over junk food.
- *Destination spa farms.* This type of microfarm is operated on the grounds of the spa itself. It allows clients to see the gardens where their healthful cuisine and herbs for body care are produced. The microfarm layout is often designed as a healing oasis, with flowers, scents and the sounds of water. The Golden Door Health Spa in Escondido, California is a fine example. It only uses the freshest ingredients from its own gardens, or neighboring organic and free-range farms.
- *Spiritual retreat centers* sometimes provide their own microfarm similar to the destination spas above.
- *Wedding host.* Farms that host weddings can offer U-pick cutting gardens, flower arranging, garden photography backgrounds, rose petals and butterflies. A butterfly farmer in Pennsylvania sells butterflies for $100 a dozen during the bridal season. Environmentalists say the release of butterflies in this manner is helping the rainforests and helping bring public awareness of the eco-system.
- *Cart farm.* This is a microfarm that sells its goods from the farmer's own cart business. Food carts—the vendors on the streets we think of as selling hot dogs and ice cream—allow food sellers to go where the people go with only a small initial investment (vs. the investment of a restaurant or bed and breakfast inn.) Fruits and veggies

already sliced and ready-to-eat can be grown on the owner's own minifarm.

- *Juicer's farms.* Local juice bars are the main target of this farm. Juice bar customers request an ongoing supply of blends of fresh, preferably organic fruits and vegetables. Carrots, beets, celery, apples, melons and berries of all kinds are a few of the better-known staples.
- *Macrobiotic farm.* This serves groups of people who adhere to a macrobiotic diet, and the vegetables called for are not always readily available. Daikon radishes and many of the cruciferous vegetables are some of the mainstays of which macrobiotic customers need a steady supply.
- *Pond and wetland nurseries.* Nurseries of this sort specialize in pond plants and unique flowering plants that enjoy moist ground, such as pussy willows, ferns, miniature cattails, primroses and mints.
- *Picnic farm.* U-picks often set up a few picnic tables, but some mini-farms with farm stands are taking it further, creating an on-farm ambience for mingling with others on their lunch hours, families who want to picnic in the country or couples or individuals seeking quiet solitude. They become a destination for eating outdoors.
- *Old-fashioned four seasons farms.* These specialize in one favorite traditional crop per season, with a favorite assortment being flowers in spring, blueberries in summer, pumpkins in fall and living Christmas trees in winter. Micro eco-farms are known for bringing the new and unusual, which includes forgotten crops that grow year-round; but they do not leave out the familiar seasonal favorites. They just do them in a special way. They grow them sustainably, allow U-picks, and perhaps have a festival for each season.
- *Christmas woodland farm.* On a living woodland Christmas farm, living Christmas trees are burlapped and set out in a woodland for customers to search through the woods for their tree. Complementary winter products include chestnuts; live evergreen trimmings including holly, ivy and cedar, and firewood (some trees such as

red alder can be "coppiced," allowing wood to be harvested and the tree to sprout back).

- *Farmer's snack stand.* Similar to roadside stands, but instead of concentrating on the raw products to be taken home and prepared, this farm grows crops especially suited to be served ready-to-eat on the spot, including fruits that can be eaten whole: apples, tomatoes, whole baby carrots, and whole lemon cucumbers, as well an assortment of prepared veggie and fruit trays.
- *Fruit and dairy deliveries.* There are still farms that sell milk in glass bottles, and home delivery is returning to favor. Fresh-picked fruit and organic dairy are a combination that may attract customers.
- *Membership farms.* Similar to CSA farms, these add the privilege of allowing the customer "members" to stroll and "hang out" and jog and take photos and just spend time there for no particular reason. The atmosphere of the farm is a selling product. Unlike the pay advance and weekly delivery as in a CSA, they pay a membership fee to be able to shop at the farm at will and to pay by the honor system.
- *Houseplant farm.* In Medford, Oregon, a mother of three children grossed close to $100,000 by growing houseplants in her backyard greenhouses. See the book *MetroFarm* in the Resources and Networking section for more details on this and other inspiring stories.
- *Organic/gourmet personal chef microfarm.* A personal chef comes to the home of his or her customers and cooks a homemade meal. They may complete a family interview if it's a family they are serving, to determine dinner favorites, discuss allergies, nutrition standards, and then they shop for all the ingredients before going to their client's homes to prepare meals. The fruits, herbs, vegetables and preserves a personal chef can offer from his or her own minifarm are endless and can make for a very unique, personalized service. One such farmer/chef simply tacked fliers on health food store and health club bulletin boards to generate customers.
- *Small goat dairy.* A special goat dairy in Nebraska thrives on five acres. The owner, Diana McCown, is a university instructor whose

family heritage included feta cheese making. She toured Greece and perfected her own cheese version. The milk is processed into cheese at a food-processing center at a local university and Diana also grows the herbs for her six different goat cheeses.

- *Sustainable hydroponics farm.* I used to think the term "sustainable hydroponics" was similar to "dry water" when high-tech and chemical formulas are used to grow food without soil. But sustainability is making its way into hydroponics farming. Twin Pine Farm is an example of a wonderful indoor oasis cultivating crops year-round in cold-weather climates. Inside their greenhouse, they allow ladybugs, lacewings, parasitic wasps and tree frogs to replace pesticides. They also keep a year-round beehive for pollination.
- *Garden shop farm.* It's been discussed already that microfarms blend very well with cottage industries and other home businesses. Backyard garden shops selling unique and handcrafted garden items along with bedding plants and arranged flowers have made several microfarmers very happy.
- *Historical farms* have been touched on in this book. There's the woman who grows historical herbs, and farms that produce heirloom vegetables and raise antique cows. My own farm (see Epilogue) added a touch of history as an experience for customers. Some farms make a certain period in time their entire farm's theme, such as colonial kitchen gardens, where crops and animals are produced and grown in similar manner and layout as they were in the pre-industrial era. Festivals and news stories attract many customers to these farms.

Some farmers choose to produce just one very traditional crop, but then offer it in many different ways. Following are two examples:

- *Potatoes only:* Pink, purple, blue and yellow-fleshed potatoes, new potatoes, fingerlings, potato gift baskets, U-dig, weekly subscription, weekly delivery to restaurants, caterers, personal chefs, bed and breakfasts. Potatoes, when fresh dug from fertile soil, are almost a different crop than mass-produced potatoes. Their skins are tender, never sprayed with anything to toughen them up or inhibit sprouting.

- *Grapes only:* Blue-black, red, purple, green, pink, table grapes, juicing grapes, jelly, juice and jam. U-pick, on-farm pressings, delivery to juice bars and bed and breakfasts. Grapes come in a huge assortment of flavors, such as "Glenora" with its spicy blueberry-like flavor, and "Einset" with a hint of strawberry flavor. Ripening times can be staggered for ongoing fresh production throughout early summer and fall.

Resources & Networking

Networking, supply sources, publications, law and policy updates, crop pricing, workshops, internships, forums, associations. Resources may combine several categories, such as organic growing associations that also offer magazines and networking.

Sustainable Organizations and Associations

Abundant Life Seed Foundation
Open-pollinated seeds, education and networking.
P.O. Box 772, Port Townsend, WA 98368, 360-385-5660
abundant@olypen.com
www.abundantlifeseed.org

Association Kokopelli
Seed saving society from France.
contactus@organicseedsonline.com
www.terredesemences.com

ATTRA (Appropriate Technology Transfer for Rural Areas)
Technical assistance, free of charge, to current and aspiring sustainable farmers in all 50 states.
P.O. Box 3657, Fayetteville, AR 72702, 800-346-9140
www.attra.ncat.org

Biodynamic Farming and Gardening Association, Inc.
25844 Butler Road , Junction City, OR 97448 , 888-516-7797 or 541-998-0105 , fax: 541-998-0106
biodynamic@aol.com

Bioneers
Solutions for restoring the earth with sustainable farm networks and workshops.
Collective Heritage Institute, 901 West San Mateo Rd., Suite L, Santa Fe, NM 87505, 877-246-6337
info@bioneers.org
www.bioneers.org

Diet For a Small Planet and Hope's Edge Organization
Frances Moore Lappe and Anna Lappe, Small Planet Fund at the Funding Exchange, 666 Broadway, Suite 500, New York, NY 10012
www.dietforasmallplanet.com

Ecology Action
Biointensive growing method.
18001 Shafer Ranch Road, Willits, CA 95490-9626, 707-459-6410
bountiful@sonic.net
www.growbiointensive.org
www.bountifulgardens.org

Homestead.org
Rural living information, networking, resources.
www.homestead.org

N.O.R.M.
National Organization for Raw Materials, Randy Cook, President, 680 E. 5 Point Highway, Charlotte, MI 48813, 517-543-0111
rccook@voyager.net
www.normeconomics.org

Organic Trade Association
P.O. Box 547, Greenfield, MA 01301, 413-774-7511
llutz@ota.com
www.ota.com

Remineralize the Earth, Inc.
Foundation dedicated to restoring all of the earth's lost minerals in original balance.
152 South St., Northampton, MA 01060-4021, 413-586-4429
reminearth@aol.com
www.remineralize-the-earth.org

Seed Savers Exchange
3076 North Winn Rd., Decorah, Iowa 52101, 319-382-5990
arllys@seedsavers.org
www.seedsavers.org

Slow Food Movement
international@slowfood.org
www.slowfood.com

The Campaign to Label Genetically Engineered Foods
P.O. Box 55699, Seattle, WA 98155, 425-771-4049
label@thecampaign.org
www.thecampaign.org

The Square Foot Gardening Foundation
info@squarefootgardening.com
www.squarefootgardening.com

Supply or Seed Sources

Acadian Seaplants Limited
Kelp meal for soil and animal rations.
30 Brown Avenue, Dartmouth, Nova Scotia, Canada B3B 1X8, 800-575-9100
info@acadian.ca
www.acadianseaplants.com

Baker Creek Heirloom Seeds
2278 Baker Creek Road, Mansfield, MO 65704, 417-924-1222
magazine@rareseeds.com
www.rareseeds.com

Caprine Supply
Products for goats, including small, rarer breeds: milk, cheese, soap, packing, more.
P. O. Box Y, De Soto, KS 66018, 913-585-1191, Fax: 913-585-1140
info@caprinesupply.com
www.caprinesupply.com

Forestfarm
A huge variety of high quality trees and shrubs from wild lilacs to native maples.
990 Tetherow Rd., Williams, OR 97544-9599, 541-846-7269
forestfarm@rvi.net
www.forestfarm.com

Garden State Heirloom Seed Society
P.O. Box 15, Delaware, NJ 07833
www.historyyoucaneat.org

Gardens Alive
Sustainable gardening products including sea and rock dust soil additions and beneficial microbes.
5100 Schenely Place, Lawrenceburg, IN 47025, 513-354-1482
www.gardensalive.com

Growing Solutions
Compost tea equipment and ingredients.
888-600-9558
www.growingsolutions.com

Heirloom Acres LLC
P.O. Box 194, New Bloomfield, Mo. 65063, 573-491-3001
www.heirloomacres.net

Heirloom Roses, Inc.
24062 NE Riverside Drive, St. Paul, OR 97137, 800-820-0465
info@heirloomroses.com
www.heirloomroses.com

Johnny's Selected Seeds
955 Benton Ave., Winslow, Maine 04901, 207-861-3900
info@johnnyseeds.com
www.johnnyseeds.com

Jon's Heirloom Plants
P.O. Box 54, Mansfield, MO 65704, 870-404-4771
jonsplants@yahoo.com
www.jonsplants.net

Lois G. Lenz
Website with source of alkalizing water and cleansers, and other natural remedies.
www.ascendingenterprises.com

Marianna's Heirloom Seeds
1955 CCC Road, Dickson, TN 37005
www.mariseeds.com

Native Seeds Search
526 North 4th Ave., Tucson, AZ 85705-8450, 520-622-5561
dpeel@nativeseeds.org
www.nativeseeds.org

Neptune's Harvest
Organic sea-based fertilizers.
88 Commercial Street, Gloucester, MA 01930, 800-259-GROW (4769)

Peaceful Valley Farm Supply
Supplies for sustainable agriculture.
P.O. Box 2209, Grass Valley, CA 95945, 888-784-1722
www.groworganic.com

Raintree Nursery
391 Butts Road, Morton, WA 98356, 360-496-6400, Fax 888-770-8358
info@raintreenursery.com
www.raintreenursery.com

Seeds of Change
Sustainably produced seeds from around the world, rare fruit trees.
P.O. Box 15700, Santa Fe, NM 87592-1500, 888-762-7333
gardener@seedsofchange.com
www.seedsofchange.com

Select Seeds

Antique flower seeds
180 Stickney Rd., Union, CT 06076, 860-684-9310
info@selectseeds.com
www.selectseeds.com

SoilSoupInc.
9792 Edmonds Way #247, Edmonds, WA 98020, 877-711-7687,
Services@soilsoup.com,
www.soilsoup.com

South Meadow Fruit Gardens
Rare and connoisseur fruit trees.
10603 Cleveland Ave., Baroda, MI 49101, 269-422-2411
smfruit@aol.com
www.southmeadowfruitgardens.com

Southern Exposure Seed Exchange
P.O. Box 460, Mineral, VA 23117, 540-894-9481
gardens@southernexposure.com
www.southernexposure.com

St. Lawrence Nurseries
Rare fruit trees.
325 State Hwy. 345, Potsdam, NY 13676, 315-265-6739
trees@sln.potsdam.ny.us
www.sln.potsdam.ny.us

The Cook's Garden
Seeds and supplies for kitchen gardeners.
P.O. Box 535, Londonderry, VT 05148, 800-457-9703
info@cooksgarden.com
www.cooksgarden.com

The Territorial Seed Company
P.O. Box 158, Cottage Grove, OR 97424-0061, 541-942-9547
tertrl@territorial-seed.com
www.territorial-seed.com

Books and Periodicals

Acres USA
A voice for eco-agriculture, monthly publication, conferences, huge selection of books which includes many hard-to-find books including *The Complete Herbal Handbook for Farm and Stable* by Juliette de Bairacli Levy and work by Andre Voisin.
P.O. Box 91299, Austin, TX 78709, 800-355-5313
info@acresusa.com
www.acresusa.com

BackHome Magazine
Small farming and sustainable living.
P.O. Box 70, Hendersonville, NC 28742, 800-992-2546
www.backhomemagazine.com

Backyard Market Gardening
By Andrew Lee
Production, marketing, for market gardeners. See *New World Publishing Bookshelf* below.

Cash from Square Foot Gardening
by Mel Bartholomew
The Square Foot Gardening Foundation
info@squarefootgardening.com
www.squarefootgardening.com

Diet For a Small Planet, and Hope's Edge
www.dietforasmallplanet.com

Four-Season Harvest
by Eliot Coleman, Kathy Bary, and Barbara Damrosch
www.fourseasonfarm.com

Growing for Market
News and ideas for market gardeners.
P.O. Box 3747, Lawrence, Kansas 66046, 800-307-8949, Fax: 785-748-0609
growing4market@earthlink.net
www.growingformarket.com

Hobby Farms Magazine
Rural living for pleasure and profit.
P.O. Box 58701, Boulder, CO 80322-8701, 800-365-4421
fancy@neodata.com

How To Grow More Vegetables Than You Ever Thought Possible on Less Land Than You Can Imagine
by John Jeavons
www.growbiointensive.org,
www.bountifulgardens.org

LaSagna Gardening
by Patricia Lanza, Rodale Press, Inc.
www.lasagnagardening.com

MetroFarm: The Guide to Growing for Big Profit on a Small Parcel of Land
by Michael Olson
P.O. Box 1244, Santa Cruz, CA 95061, 831-427-1620
michaelo@metrofarm.com
www.metrofarm.com

Michael McGroarty
Books and articles on growing landscape plants on 1/20th of an acre.
P.O. Box 338, Perry, Ohio 44081
mcplants@ncweb.com
www.freeplants.com

New World Publishing
Publisher of this book, as well as *Sell What You Sow!, The New Farmers' Market,* and *Wild Herbs.* Online catalog, *New World Publishing Bookshelf,* has dozens of titles on sustainable agriculture, green building, and renewable energy. (See p. 175).
11543 Quartz Dr. #1, Auburn, CA 95602, 530-823-3886
nwpub@nwpub.net
www.nwpub.net

Organic Gardening Magazine
33 East Minor Street, Emmaus, PA 18098, 610-967-5171
OGDcustserv@cdsfulfillment.com
www.organicgardening.com

How to Have a Green Thumb Without an Aching Back: A New Method of Mulch Gardening
by Ruth Stout
www.earthlypursuits.com

Sharing The Harvest
By Elizabeth Henderson
All about Community Supported Agriculture (CSA). See *New World Publishing Bookshelf* above.

Solar Gardening: Growing Vegetables Year-Round the American Intensive Way (American Intensive Method)
by Leandre Poisson and Gretchen Vogel Poisson
Chelsea Green Publishing
www.chelseagreen.com

The Growing Edge Magazine
Hydroponics, aeroponics, greenhouses, explores sustainable methods.
P.O. Box 1027, Corvallis, OR 97339-1027, USA: 800-888-6785, Worldwide: 1-541-757-8477, Fax: 541-757-0028
www.growingedge.com

The Heirloom Gardener Magazine
Hill Folk's Publishing, 2278 Baker Creek Road, Mansfield, MO 65704, 417-924-1222
magazine@rareseeds.com
www.theheirloomgardener.com

The Permaculture Activist
Publication for Permaculture, the study of allowing the earth to do more and more, the human to do less and less, while increasing productivity.
P.O. Box 1209, Black Mountain, NC 28711, 828-669-6336
pcactivist@mindspring.com
www.permacultureactivist.net
Other sources for Permaculture information include: *Acres USA* (see above).

Networking, Workshops, Information Centers, etc.

Andre Voisin (see Acres)

CSA-L
E-mail list and discussion group on Community Supported Agriculture (CSA)
http://www.prairienet.org/pcsa/CSA-L

GardenWeb Forums
The largest community of gardeners on the internet, with many threads of interest to high-value crop growers and marketers.
http://forums.gardenweb.com/forums

Ken Hargesheimer
Workshops and information on organic, biointensive, raised-bed gardening, market gardening, mini-farming, and mini-ranching worldwide.
P.O. Box 1901, Lubbock, Texas 79408-1901, 806-744-8517, Fax 806-747-0500
minifarms@aol.com
www.minifarms.com

Market Farming
E-mail list and discussion group plus helpful articles on market farming.
www.marketfarming.net

Organic Research.com
Updates on policies, certification, organic farming research.
CABI Publishing North America, 44 Brattle Street, 4th Floor, Cambridge, MA 02138, 800-528 4841
tbrainerd@pcgplus.com

Peter Weis, Futurist
Holistic health, sustainable farming, restoring the 72+ trace elements.
#306 - 1035 Pendergast Street, Victoria, BC, Canada V8V 2W9
pweis@direct.ca
www.truehealth.org

Sylvia and Walter Ehrhardt
Internships and information on small organic farming.
Fax: 301-834-5070
ecoag@igc.apc.org

The "NEW FARM" web site
Information on organic farming, price indexes for organic foods, and forum for organic farmers.
www.newfarm.org

The Stewardship Community
Guidestone CSA Farm and Center for Sustainable Living, 5943 N. County Road 29, Loveland, Colorado 80538, 970-461-0272
guidestone_alliance@msn.com
www.stewardshipcommunity.org

Index

New World Publishing

Thank you for purchasing *Micro Eco-Farming!* If you like it, please tell others! Quantity order discounts: See *Orders* below.

Sell What You Sow! The Grower's Guide to Successful Produce Marketing, by Eric Gibson. This book was described by *Small Farmer's Journal* as "far-and-away the outstanding farm produce marketing text." Deciding what to grow, selling through farmers markets, roadside markets, U-pick, Community Supported Agriculture (CSA), mail-order, grocery stores and restaurants, processed food products, advertising and promotion, rules, regulations and insurance, pricing and more. 8 1/2 x 11, 304 pps. $24.95.

The New Farmers' Market: Farm-Fresh Ideas for Producers, Managers & Communities, by Vance Corum, Marcie Rosenzweig & Eric Gibson. Definitive book on farmers' markets for farmers, market managers or city planners. 8 x 10, 272 pps. $24.95.

New World Publishing Bookshelf: Online catalog includes books on market gardening, sustainable agriculture, renewable energy, alternative ("green") building, social change. See website (below).

Free Online Downloads: *The Hot 50 Marketing Tips, The Vegetarian Alternative* and others. See website (below).

Orders & Information: Single orders: Call 888-281-5170 or online at website (below). Quantity orders: Call 530-823-3886.

New World Publishing, 11543 Quartz Dr. #1
Auburn, CA 95602. Ph. & Fax: 530-823-3886
nwpub@nwpub.net • www.nwpub.net

About the Author

Drawing from experience as a micro eco-farmer on her small island American farm, interaction with hundreds of others in the field, and years of research on the topic, Barbara Berst Adams now writes feature articles for national magazines on micro eco-farming, small enterprise, and creating a sustainable universe. She continues her research and consults on abundant microfarm techniques on a growing number of continents: from North American communities to African orphan villages.

Barbara invites you to visit her website:

www.MicroEcoFarming.com

for updated articles, more successful micro eco-farm profiles from around the world, color photographs and resources for schools.